Emergency/Standby Power Systems

Emergency/Standby Power Systems

Alexander Kusko

Life Fellow, Institute of Electrical and Electronics Engineers
Lecturer in Electrical Engineering,
Massachusetts Institute of Technology
Director, Kusko Electrical Division
Failure Analysis Associates

McGraw-Hill Book Company

New York   St. Louis   San Francisco   Auckland
Bogotá   Hamburg   London   Madrid   Mexico
Milan   Montreal   New Delhi   Panama
Paris   São Paulo   Singapore
Sydney   Tokyo   Toronto

Library of Congress Cataloging-in-Publication Data

Kusko, Alexander, date
    Emergency/standby power systems.

    Includes index.
    1. Emergency power supply   I. Title.
TK1005.K87   1989   621.31   88-12856
ISBN 0-07-035689-0

1234567890   DOC/DOC   89321098

ISBN 0-07-035689-0

The editors for this book were Harold B. Crawford and Galen H. Fleck, the designer was Naomi Auerbach, and the production supervisor was Richard A. Ausburn. This book was set in Century Schoolbook. It was composed by the McGraw-Hill Book Company Professional & Reference Division composition unit.

Printed and bound by R. R. Donnelley & Sons Company.

Contents

Preface   xi

Part 1   Introduction and Basic Systems

Chapter 1. Introduction   3

    1.1   Purpose   4
    1.2   History   4
    1.3   Load Sensitivity   6
    1.4   Definitions   6
    1.5   Outline of the Book   7
        References   8

Chapter 2. Basic Systems   9

    2.1   Storage Battery—Emergency DC System   9
    2.2   Engine-Generator System   11
    2.3   UPS System   12
    2.4   Separate Service   14
    2.5   Connection Ahead of Service-Disconnecting Means   15
    2.6   Unit Equipment   15
    2.7   Summary   16
        References   16

Part 2   Equipment

Chapter 3. Motor-Generator Set   19

    3.1   Standards   19
    3.2   Synchronous Generator   20
    3.3   Synchronous/Synchronous MG Set   24
    3.4   Induction/Synchronous Set   25
    3.5   Ride-Through Capability   27
    3.6   Uninterruptible MG Set   28
    3.7   Summary   30
        References   31

Chapter 4. Transfer Switches   33

4.1   Codes and Standards   33
4.2   Types of Transfer Switches   34
4.3   Construction   35
4.4   Operation   37
4.5   Controls   41
4.6   Static Transfer Switch   44
4.7   Summary   47
      References   48

Chapter 5. Engine-Generator Sets   49

5.1   Standards   49
5.2   Types   50
5.3   Generators   51
5.4   Generator Voltage Waveform   54
5.5   Generator Loading   57
5.6   Generator Protection   60
5.7   Engines   64
5.8   Controls   66
5.9   Installation   68
5.10 Summary   70
      References   70

Chapter 6. Static UPS   73

6.1   Standards   74
6.2   Application   74
6.3   UPS Module   75
6.4   Theory of Operation: Rectifier   77
6.5   Theory of Operation: Inverter   78
6.6   Nonlinear Load   85
6.7   UPS Control Section   89
6.8   Reliability Improvement   91
6.9   Installation   91
6.10 Summary   97
      References   98

Chapter 7. Batteries   99

7.1   Codes and Standards   99
7.2   Electrochemistry of Batteries   100
7.3   Types of Batteries   102
7.4   Characteristic Curves   106
7.5   Sizing   108
7.6   Maintenance   112
7.7   Summary   116
      References   117

Chapter 8. Power Distribution Units 119

8.1 Components of the Power Distribution Unit   119
8.2 Power Distribution Unit   120
8.3 Voltage-Regulating Transformers   121
8.4 Isolation Transformer   123
8.5 Standard Connections   126
8.6 Summary   126
     References   127

Part 3 Examples of Emergency/Standby Systems in Use

Chapter 9. Computer Centers 131

9.1 Terminology   132
9.2 References   132
9.3 Typical System   133
9.4 Example: Affiliated Food Stores   134
9.5 Example: Federal Reserve Board   135
9.6 Example: Wakefern Food Corp.   137
9.7 Example: Amoco Computer Center   141
9.8 Summary   142
     References   144

Chapter 10. Health Care Facilities 145

10.1 Standards   145
10.2 Definitions   146
10.3 Example: Typical Hospital Wiring Arrangement   148
10.4 Example: Nursing Home Wiring Arrangement   149
10.5 Example: Hospital System   149
10.6 Summary   151
      References   151

Chapter 11. Office Buildings 153

11.1 Building Loads   153
11.2 Example: PPG Headquarters   155
11.3 Example: Dow Jones Offices   157
11.4 Summary   162
      References   162

Chapter 12. Remote Sites 163

12.1 Types of Electric Power Supplies   163
12.2 Example: Photovoltaic Power System for Telecommunications   166
12.3 Example: Repeater for Optical Fiber Cable   167
12.4 Example: Small Earth Station for Satellite   167
12.5 Summary   173
      References   173

Part 4   Procedures

Chapter 13. Load Classification   177

13.1   Utility Power   177
13.2   Alternate Power Sources   179
13.3   Load Categories   180
13.4   Loads for Battery Supply   181
13.5   Loads for UPSs   182
13.6   Loads for Engine-Generator Sets   182
13.7   Summary   185
       References   185

Chapter 14. Reliability   187

14.1   Definitions   187
14.2   Sources of Reliability Data   188
14.3   Reliability Essentials   190
14.4   Reliability Calculation: Single Module   191
14.5   Reliability Calculation: Redundant Systems   193
14.6   Static Bypass Switch   194
14.7   Summary   196
       References   196

Chapter 15. Installation   199

15.1   Types of Equipment   199
15.2   Equipment Locations   200
15.3   Circuits   200
15.4   Protection   204
15.5   Generator Protection   205
15.6   Grounding   206
15.7   Maintenance   209
15.8   Summary   210
       References   210

Chapter 16. Procurement   213

16.1   Procurement Process   213
16.2   Engineering   214
16.3   Specifications   214
16.4   Purchasing   215
16.5   Testing   217
16.6   Operation and Maintenance   218
16.7   Summary   218
       References   218

Chapter 17. Cost/Benefit Analysis   219

17.1 Frequency of Utility Power Failure   220
17.2 Cost of Utility Failures   220
17.3 Summary   226
      References   226

Part 5   Codes and Standards

**Chapter 18. Codes Governing Emergency/Standby Electric
            Power Systems**   229

18.1 ANSI/NFPA 110-1985, "Standard for Emergency and
      Standby Power Systems"   230
18.2 ANSI/NFPA 70-1987, "National Electrical Code®"   230
18.3 ANSI/NFPA 99-1984, "Essential Electrical Systems for
      Health Care Facilities"   233
18.4 Summary   234
      References   234

Index   237

Preface

In the face of possible failures of normal utility electric power sources, a reliable supply of electric power must be provided for such facilities as patient health care centers, data processing systems, and critical telecommunications links. To answer the need, emergency/standby systems using batteries, inverters, engine-generator sets, and indeed the whole spectrum of electric energy sources have been developed. The installed systems range from 10-W thermoelectric generators at remote sites to multimegawatt complexes using engine-generators, switchgear, batteries, and inverters at large computer centers.

This book draws on the extensive consulting experience of the author in the emergency/standby electric power field. The content was designed for professional engineering seminars in the field. However, the book will serve the reader in all aspects of the industry, including descriptions of equipment, examples of emergency systems, and codes, standards, and references. The book is of value to engineers designing emergency/standby systems for computer centers, office buildings, health care facilities, remote communications sites, and military facilities. It will also serve the operators of such facilities, manufacturers of the equipment, and vendors of data processing equipment that must operate from the emergency/standby power. In effect, the book will serve all engineers, purchasing agents, specification writers, operators, and manufacturers who deal with equipment that either provides emergency/standby power or uses that power.

The book is organized into five parts. Part 1, the introduction, explains the components, operation, and terminology of the basic emergency system. Part 2 describes the equipment used in the systems, including motor-generator sets, batteries, uninterruptible power sources, transfer switches, power distribution units, and engine-generators. Part 3 gives examples of systems in the health care industry, computer centers, and office buildings and at remote sites. Part 4 describes procedures for analyzing, specifying, and purchasing the systems, and Part 5 deals with codes and standards.

Numerical examples are included to assist the reader in designing systems and to use in teaching a course with this book as a text.

The author wishes to express his appreciation to Mrs. Lillian Kaplan for her work in preparing the manuscript.

Alexander Kusko

Introduction and Basic Systems

Chapter

1

Introduction

Emergency/standby systems provide electric power for critical functions and equipment when the quality of the normal supply, e.g., utility power, is not adequate or fails entirely. The rapid growth in the use and installation of data processing equipment, life care and other medical equipment, alarm systems, and safety lighting has prompted a new industry of emergency/standby systems. The industry includes manufacturers of engine-generator sets, uninterruptible power systems (UPSs), motor-generator sets (MGs), and transfer switches, as well as consultants and architect/engineer (A/E) firms specializing in the design of such systems. In addition, a group of standards that define the requirements of such systems has emerged. There is hardly a facility, from the New York Stock Exchange to the local school, that does not include an emergency/standby system of some type.

An emergency power system is defined as [1] (©1984-IEEE):

> An independent reserve source of electric energy which, upon failure or outage of the normal source, automatically provides reliable electric power within a specified time to critical devices and equipment whose *failure to operate satisfactorily would jeopardize the health and safety of personnel or result in damage to property.* [Emphasis added.]

A standby power system is defined by IEEE as:

> An independent reserve source of electric energy which, upon failure or outage of the normal source, provides electric power of acceptable quality and quantity so that *the user's facilities may continue in satisfactory operation.* [Emphasis added.]

Because the definitions are so close and the same types of equipment are used in both systems, we will use the term "emergency/ standby" for all systems that operate from an independent reserve source of electric energy upon failure or outage of the normal source.

3

Likewise, we will use the term "emergency loads" for all equipment supplied from the emergency/standby system, unless defined otherwise for a specific system.

1.1 Purpose

Utility electric power systems have evolved to supply bulk electric energy to customers for loads such as lighting, heating, motor-driven appliances, and other equipment that can tolerate momentary and longer interruptions without damage and with incidental inconvenience. To supply such loads, to meet their concept of reliability, utilities employ feeder and capacitor switching, step feeder regulators, voltage reduction measures (brownouts), and occasional complete interruption for critical maintenance work. However, a small fraction of the total customer load, consisting of emergency lighting, medical facilities, data processing, and communications centers, cannot tolerate utility quality power. As a result, a family of equipment, termed "emergency/standby systems," has been developed to provide high-quality power for that portion of the customer load which requires it [2].

This book is intended for people who specify, purchase, install, operate, and maintain emergency/standby systems. The equipment used is common to a wide variety of applications, yet the terminology, the codes and standards, and the practices may differ. This book will bring together the field in one reference so that it will be more easily understandable. It is intended not as a design manual, but as a means of understanding the design problems and knowing where to obtain detailed design and performance information.

1.2 History

The development of emergency/standby systems paralleled the increased dependence of society on electric power to provide motive power and intelligence for all aspects of modern life. These systems have become necessary since the 1950s to provide emergency power for lighting and elevators in buildings, to ensure continued operation of medical equipment, to keep electronic data processing facilities in operation, and to operate telecommunications equipment at remote sites.

Standby engine-generator sets date to the early 1920s. Figure 1.1 shows an early set which Admiral Byrd took to Antarctica in 1928. These were not the quick-start sets with power quality sensors and automatic transfer switches in use today. Gasoline and diesel engine-generator sets were widely used during World War II for electric power supply at all overseas facilities. However, N. B. Tharp, as early

Figure 1.1 Engine-generator set taken to Antarctica by Admiral Byrd in 1928. (*Courtesy Kohler Co.*)

as 1955 [3], discussed the reliability of standby engine-generator and transfer switch systems rated up to 25 kW for microwave communications systems.

Storage batteries have provided floating standby power in telephone central offices back to the turn of the century. The start of the computer age in the 1950s saw the development of uninterruptible power systems (UPSs) using motor-generator sets and, later, solid-state rectifier/inverter modules. J. J. Gano in 1956 [4] reported on the power systems for the SAGE air-defense system, which employed vacuum-tube-type digital computers. In 1963, J. L. Fink, J. F. Johnston, and F. C. Krings [5] described solid-state rectifier/inverter systems for essential loads up to 100 kVA as built and up to 800 kVA on order. A. Kusko and F. E. Gilmore [6], in 1967, described the modular concept of the UPS as it was applied to the installation of large systems installed in more than twenty FAA Air Route Traffic Control Centers in the United States. Since those dates, the performance of all emergency/standby equipment has been improved constantly in reliability by using solid-state controls, microprocessors, more reliable devices, and better overall engineering and manufacturing techniques. A modern UPS installation is shown in Fig. 1.2.

The annual business of UPSs alone was estimated as $1.5 billion worldwide in 1988 and is projected as $2 billion by 1990. However, the value of the computers and other critical equipment protected by UPSs is probably ten to one hundred times the cost of the UPSs themselves.

Figure 1.2   Modern installation of a 60- and 415-Hz UPS at a computer center. (*Courtesy Emerson Electric Co.*)

1.3.   Load Sensitivity

Emergency/standby systems are required to provide electric power to load equipment, when the normal supply fails, in sufficient time to prevent jeopardizing the health and safety of personnel, prevent damage to property, and/or to ensure continued satisfactory operation of facilities. The time interval is difficult to define or measure; it is frequently set by the best performance of the available emergency/standby systems. In general, the load equipment is categorized by the need for it: (1) power is required within one-half cycle (of 60 Hz) as delivered by UPS, e.g., for computers, (2) power is required within 10 s as delivered by engine-generator sets, e.g., for fans, pumps, and emergency lighting, (3) power is required within minutes, as delivered by manually operated standby or transfer equipment, e.g., for industrial processes.

1.4   Definitions

At least four authoritative standards provide key definitions for emergency/standby systems: ANSI/NFPA 110 [7], ANSI/NFPA 99 [8], ANSI/IEEE Std. 446 [9], ANSI/NFPA 70 [10]. ANSI/NFPA 110, "Standard for Emergency and Standby Power Systems," includes the most comprehensive definitions of the supply system, as follows:*

---

*Reprinted with permission from NFPA 110-1985, Emergency and Standby Power

1. *Function.* The function of the Emergency Power Supply System (EPSS) is to provide a source of electric power of required capacity, reliability and quality for a given length of time to loads within a specified time after loss or failure of the normal supply.

2. *System.* The EPSS consists of an energy source, which may include a means to convert the energy to electric power, a system of conductors, disconnecting means, transfer switches, and all control and support devices up to and including the load terminals of the transfer equipment needed for this system to function in a safe and reliable way.

3. *Classification of EPSS.* The terms Emergency Power Supply Systems and Standby Power Supply Systems include other terms such as Alternate Power Systems, Standby Power Systems, Legally Required Standby Systems, Alternate Power Sources, and other similar terms.

The classifications of EPSS include the following:

Type. The maximum time in seconds that the EPSS will permit the load terminals of the transfer switch to be without acceptable electric power, e.g., Type U, uninterruptible UPS; Type 10, 10 s; Type M, manual transfer.

Class. The minimum time in hours the EPSS is designed to operate its rated load without being refueled, e.g., Class 0.083, 0.083 h or 5 min.; class 48, 48 h.

Category. Category A includes stored energy devices receiving their energy solely from the normal supply. Category B includes all devices not included in Category A.

Level. Three levels of equipment installation, performance, and maintenance are defined. Level 1 defines equipment performance requirements for applications where the requirements are most stringent and where failure of the EPSS equipment to perform could result in loss of human life or serious injuries. Level 2 defines equipment where failure to perform is less critical to human life and safety. Level 3 defines all other equipment and applications, including optional standby systems, not defined in Levels 1 and 2.

1.5 Outline of the Book

This book is arranged in five parts:

Part 1 Introduction. Describes functions and definitions of systems. Outlines the basic systems using combinations of batteries, inverters, motor-generators, and engine-generator sets.

Part 2 Equipment. Describes the theory, operation, and characteristics of each type of equipment found in emergency/standby systems, e.g., engine-generators, UPSs, transfer switches, and power distribution centers.

Part 3   Examples of Emergency/Standby Systems in Use. Describes the special requirements of four types of applications: health care facilities, office buildings, remote sites, and data processing centers. Descriptions of the equipment and electrical layout of specific facilities are given.

Part 4   Procedures. Describes methods for classifying the loads and calculating the reliability of emergency/standby systems and purchasing and installing the systems.

Part 5   Codes and Standards. Describes and lists pertinent standards published by such organizations as the American National Standards Institute (ANSI) and National Fire Prevention Association (NFPA).

REFERENCES

1. ANSI/IEEE Std. 100-1984, "IEEE Standard Dictionary of Electrical and Electronic Terms."
2. A. Kusko, "The Quality of Electric Power," *IEEE Trans. Ind. and Genl. Appl.,* vol. IGA-3, no. 6, November/December 1967, pp. 521–524.
3. N. B. Tharp, "Recommendations for Improving Reliability of Standby Engine Generators for Microwave Communications Systems," *Trans. AIEE,* vol. 74, pt. III, April 1955, pp. 261–267.
4. J. J. Gano, "Power Generation and Distribution at Air-Defense Computing Centers," *Elec. Eng.,* vol. 75, December 1956, pp. 1098–1102.
5. J. L. Fink, J. F. Johnston, and F. C. Krings, "The Application of Static Inverters for Essential Loads," *IEEE Trans. PAS,* vol. 82, December 1963, pp. 1068–1072.
6. A. Kusko and F. E. Gilmore, "Concept of a Modular Static Uninterruptible Power System," *Conf. Record of the 1967 IEEE Industry and General Applications Group Annual Meeting,* IEEE 34C62, pp. 147–153.
7. ANSI/NFPA 110-1985, "Standard for Emergency and Standby Power Systems."
8. ANSI/NFPA 99-1984, *Standard for Health Care Facilities,* Chap. 8, "Essential Electrical Systems for Health Care Facilities."
9. ANSI/IEEE Std. 446-1987, "IEEE Recommended Practices for Emergency and Standby Power Systems for Industrial and Commercial Applications" (Orange Book).
10. ANSI/NFPA 70-1987, "National Electrical Code®."*

---

*National Electrical Code® and NEC® are registered trademarks of the National Fire Protection Association, Inc., Quincy, Mass.

Basic Systems

Emergency/standby systems are categorized by the sources of electric power. The design of a system depends upon the duration of operation, the power required, the reliability, and the time to transfer from the normal source to the alternate source.

The sources of power for the basic systems are listed in Art. 700-12, ANSI/NFPA-70, "National Electrical Code®" (NEC®) [1]:*

1. Storage battery

2. Engine-generator set

3. Uninterruptible power supply (UPS)

4. Separate service

5. Connection ahead of service-disconnecting means

6. Unit equipment (emergency lighting)

Each of these systems will be explained in this chapter.

Each source of power is integrated into a basic (sub)system that is used alone, or combined with others, to form a legally or nonlegally required emergency/standby system for a particular site. The individual systems will be described briefly in this chapter and in more detail in later chapters.

2.1 Storage Battery—Emergency DC System

A one-line diagram for an emergency dc system is shown in Fig. 2.1. The system consists of an automatic charger, a battery, and an emer-

---

*National Electrical Code® and NEC® are registered trademarks of the National Fire Protection Association, Inc., Quincy, Mass.

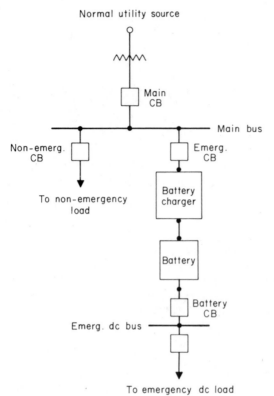

Normal utility source

Main CB

Main bus

Non-emerg. CB

To non-emergency load

Emerg. CB

Battery charger

Battery

Battery CB

Emerg. dc bus

To emergency dc load

Figure 2.1 Emergency dc system with a battery source.

gency dc bus. It is used primarily for emergency lighting, fire alarms, and emergency communications systems.

The NEC® requires that the system be capable of maintaining the total load for a period of 1½ h minimum without the load voltage falling below 87.5 percent of normal. The acid or alkali batteries shall be suitable for emergency service and be compatible with the charger. Automotive batteries are not permitted.

The concept of a charger and a battery to provide a reliable dc power supply that is independent of the utility line is widely used in non-legally-required applications. For example, telephone central offices depend upon large battery banks to provide an electrically noise-free, highly reliable power source for subscribers' telephone services. Utility substations and power plants employ battery banks to provide dc power for protective relays, tripping and closing circuit breakers, operating backup turbine-generator lubrication pumps, and various indicator lamps and annunciators.

2.2 Engine-Generator System

The most common standby power source is an engine-generator set. The set consists of a gasoline or diesel engine or a gas turbine, an ac synchronous generator, and the controls. In standby service, the set is always used with a transfer switch arrangement. Although the set is the ideal alternate power source after it has been started and reaches operating speed, it requires considerable supplementary equipment to install it and a well-organized program to maintain it. For example, the set requires a foundation to carry its weight, a fuel system, an exhaust system, a ventilation system, controls, and switchgear. Gasoline engines, diesel engines, and gas turbines are employed as the power ratings of the sets are increased.

A one-line diagram for a system in which an engine-generator provides the alternate power source is shown in Fig. 2.2. The load is supplied from the normal utility source. When the normal source goes out of voltage and/or frequency limits or fails completely, the engine is started. As soon as the generator output reaches nominal voltage and frequency, the transfer switch operates to place the emergency load on the generator. When the normal power returns with voltage and frequency within limits for a prescribed time, the transfer switch is op-

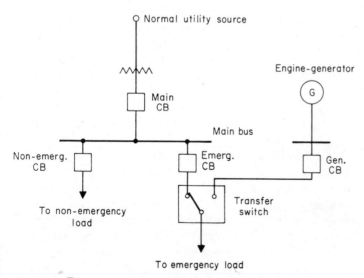

Figure 2.2 Emergency engine-generator system with a transfer switch.

erated manually or automatically to return the emergency load to the normal supply.

The transfer switch manufacturer provides the devices that sense the conditions of the voltage and frequency of the normal supply line. When one condition goes out of the acceptable range, the device operates to start the engine-generator set, detect readiness of the set, and cause the transfer switch to operate. The reverse sequence occurs when the normal supply line returns to operation.

The time for an engine to start and for the load to be transferred to the generator is about 10 s. During the starting time, the emergency loads are without power. The loads may be reapplied to the generator sequentially to prevent the starting or inrush current from causing extraneous voltage dips. In contrast to a UPS, the engine-generator allows a 10-s zero-power period. Unlike batteries, however, the engine-generator can operate for hours, provided that fuel is supplied.

The NEC® sets numerous requirements for legally required engine-generator sets. For example, the set shall not transfer its emergency load back to the normal source in less than 15 min. The on-premise fuel supply shall provide for at least 2 h of full-demand operation. The engines shall not be solely dependent upon a public utility gas system for fuel or on a municipal water system for cooling. Obviously, the system shown in Fig. 2.2 can be built with multiple paralleled generators and multiple transfer switches for supplying emergency loads at Levels 1, 2 and 3 as defined in Sec. 1.4.

2.3 UPS System

A one-line diagram for a system in which a static (solid-state) UPS provides the emergency power is shown in Fig. 2.3. The UPS module consists of a battery charger, a battery, and an inverter. The module is usually provided with a bypass circuit that transfers the emergency load to the normal source either automatically if the UPS fails or manually to isolate the UPS for maintenance. The high-speed static switch closes first, followed by the closing of the bypass circuit breaker (CB) and opening of the output CB. The UPS operates to supply the emergency load continuously; it does not operate in a standby mode.

UPS modules up to about 750 kVA are available. Modules are also built for 60-Hz input and 415-Hz output to supply specific types of computers. Modules are usually installed in multiple parallel configurations to supply large emergency loads and to increase the system reliability. With an extra or redundant module in the system, any one module can be removed from service should it fail or have to be maintained, and the remaining modules can carry the load.

The battery in a UPS system is typically sized to carry the emer-

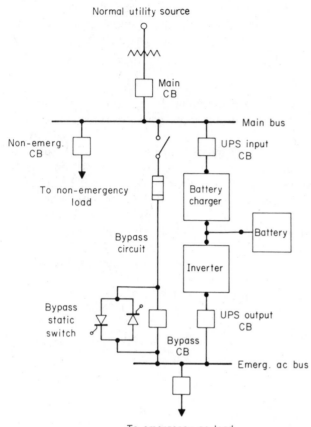

Normal utility source

Main
CB

Main bus

Non-emerg.
CB

To non-emergency
load

UPS input
CB

Battery
charger

Battery

Bypass
circuit

Inverter

Bypass
static
switch

UPS output
CB

Bypass
CB

Emerg. ac bus

To emergency ac load

Figure 2.3   Emergency system with a static UPS.

gency load for 5 to 20 min. An engine-generator system, such as the one described in Sec. 2.2, is frequently combined with a UPS system to extend the running time of the UPS beyond the battery time when the normal source fails. Not only can the engine-generator system supply the UPS modules, it can carry Level 3 emergency loads as well.

Instead of a static UPS module, a motor-generator (MG) set can be used to isolate an emergency load from the normal supply. An MG set can supply an emergency load from its kinetic energy for only about 100 ms when the normal supply fails completely. Various modifications can be used to extend the operating time when the normal supply fails. For example, a flywheel can be added to extend the time to several seconds. It can be supplemented by an engine and clutch to pick up the set before its output frequency and voltage fall below limits. Finally, the set can be designed as a rotary uninterruptible power supply, called an RUPS, by using a battery as the alternate energy supply. The battery provides power either to an auxiliary dc motor on the shaft or to a static inverter which powers the ac drive motor.

2.4 Separate Service

A one-line diagram for a system in which the emergency load can be supplied from the main bus or an alternate utility service is shown in Fig. 2.4.

The load is normally supplied by the normal supply line from the utility. If the normal supply fails, the transfer switch automatically transfers the emergency load to the alternate supply line. To be effective, the alternate line should come by a different route, and even from a different substation, than the normal line. Successful operation depends upon the correct operation of the voltage-monitoring relays in the transfer switch. Actual transfer will occur in less than 1 s after relay operation. The emergency load can be returned to the normal line manually or automatically.

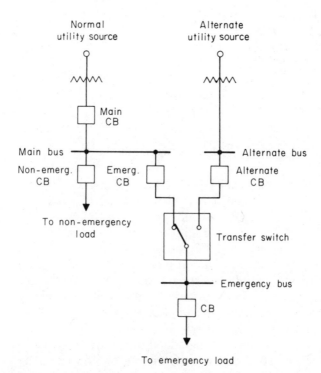

Figure 2.4   Emergency system with an alternate utility service.

Transfer to the alternate line cannot be sufficiently fast to prevent the shutdown of computers, data processing equipment, motors on undervoltage trip, and high-pressure mercury lamps. All will have to be restarted. Furthermore, the blackout-type area loss of all utility power will terminate service on both lines. The alternate line system is suitable for nonessential commercial and industrial emergency loads, but the electronic portion of the loads will require a more secure supply, such as a UPS.

Variations of the system include (1) the two lines paralleled through two circuit breakers to the main bus or (2) the two lines connected through circuit breakers to two buses with a third tie breaker between the buses.

Separate service, particularly by a different route or from a different substation than the normal service is not always feasible and is always costly. It is justifiable only for large industrial loads and large building complexes that utilize most of the capacity of either feeder. A commonly used alternative is to supply the site with two feeders in a loop system. One transformer is employed; the primary side can be supplied by a feeder from either direction in the loop.

2.5   Connection Ahead of Service-Disconnecting Means

From Art. 700-12(e) [1]:*

> Where acceptable to the authority having jurisdiction, connections ahead of, but not within, the main service disconnecting means shall be permitted. The emergency service shall be sufficiently separated from the normal main service disconnecting means to prevent simultaneous interruption of supply through an occurrence within the building or groups of buildings served.

2.6   Unit Equipment

From Art. 700-12(f) [1]:†

> Individual unit equipment for emergency illumination shall consist of: (1) a rechargeable battery; (2) a battery charging means; (3) provisions for one or more lamps mounted on the equipment and/or shall be permitted to have terminals for remote lamps; and (4) a relaying device ar-

---

*Reprinted with permission from NFPA 70-1987, National Electrical Code®, Copyright© 1986, National Fire Protection Association, Quincy, MA 02269. This reprinted material is not the complete and official position of the NFPA on the referenced subject which is represented only by the standard in its entirety.

†Ibid.

ranged to energize the lamps automatically upon failure of the supply to the unit equipment.

2.7 Summary

Emergency/standby systems are assembled from the basic equipment components of transfer switches, engine-generator sets, motor-generator sets, and UPS modules. These components are interconnected in various configurations to ensure reliable power supply to the load in the face of loss of utility power or failure of the equipment itself and within the requirements of the load.

REFERENCES

1. ANSI/NFPA 70-1987, "National Electrical Code®."

Equipment

3

Motor-Generator Set

As the name implies, a motor-generator (MG) set consists of an electric motor driving an electric generator. The set also includes a motor starter and a generator voltage regulator. For emergency/standby applications, the generator is a synchronous machine; the motor may be synchronous, induction, or direct current. Other than its own rotational kinetic energy, the MG set has no inherent stored energy to supply power for more than 100 ms when the normal supply fails.

An MG set can be used for the following purposes:

1. To buffer the load from voltage transients and voltage waveform distortion introduced by the normal power source.

2. To provide short-time energy storage by means of a flywheel for supply of uninterrupted power to the load until a standby source is engaged.

3. To change the frequency of the normal source voltage for specific loads, e.g., 60 to 415 Hz for computers or 50 to 60 Hz for U.S. equipment operated overseas.

Note that the MG set is in direct competition with static UPS equipment for all of the above-listed applications.

A 75-kVA MG set used to convert 60-Hz to 415-Hz power is shown in Fig. 3.1. The induction drive motor and synchronous generator are mounted on a two-bearing horizontal shaft. The MG set shown in Fig. 3.2 includes three machines on the same shaft. The main section consists of an ac drive motor and a 62-kVA ac generator. The third unit is a dc machine that serves as a battery-charging generator or a dc-drive motor for the set when the utility power fails.

3.1 Standards

The basic standard for motors and generators is ANSI/NEMA MG1 [1]. It provides practical information concerning performance, dimen-

(a)

(b)

Figure 3.1 MG set, 75 kVA, 60/415 Hz. (*a*) Standard package in front; silent block in rear. (*b*) Standard package with control compartment open. (*Courtesy K/W Control Systems*)

sions, testing, and ratings. For installation and operation, refer to ANSI/NFPA 70 [2], Art. 430 on motors, motor circuits, and controllers and Art. 445 on generators.

3.2 Synchronous Generator

Synchronous generators, also termed "alternators," are built for emergency/standby service in the range from 50 to 2500 kW, to be driven by motors or engines. Generators up to 15,000 kW are built for emergency service in nuclear power plants and for "black-starting" fossil-fueled power plant turbine-generator units under emergency conditions. The basic parts and the principle of operation are the same for generators over the cited power range. The construction and oper-

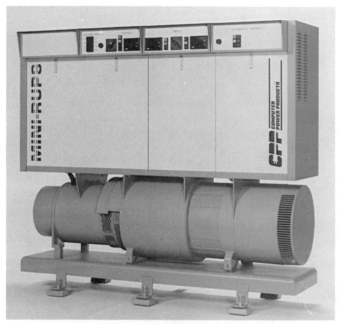

Figure 3.2   62-kVA rotary uninterruptible power supply (RUPS); dc and ac drive motors, 60-Hz ac generator. (*Courtesy Computer Power Products*)

ation of synchronous generators are described in the book by Fitzgerald, Kingsley, and Umans [3].

Construction

The synchronous generator is constructed with three functional parts: (1) the mechanical part consisting of the frame, shaft, bearings, and housing, (2) the magnetic part consisting of the ferromagnetic material in the stator and the rotor, and (3) the windings on the stator and the rotor. The stator windings deliver the output power. The rotor windings serve to establish the magnetic field in the generator and control the output voltage.

The stator and rotor of a synchronous generator are shown in Fig. 3.3. The stator consists of a stack of cylindrical steel laminations with axial teeth and slots on the inner circumference. The stator windings are placed in the slots and interconnected to form the stator terminals. The rotor consists of a structure with projecting steel poles. Coils called field windings are placed on the poles. The elementary cross-sectional diagram of a three-phase synchronous generator is shown in Fig. 3.4. The generator consists of a rotor, also called the field, and a

Figure 3.3 Disassembled four-pole synchronous generator showing stator, salient-pole rotor, and brushless exciter. (*Courtesy Katolite Corp.*)

stator. When direct current I_f is applied to the field windings, the poles become magnetized with north (N) and south (S) poles. The rotor shown in Fig. 3.4a has two poles.

Operation

For the two-pole generator of Fig. 3.4a, the stator coil sides such as a-a are spaced 180°. The successive coils such as a-a, b-b, and c-c are spaced 120° apart to obtain three-phase voltages. The a-phase winding, instead of being a single coil, is usually distributed in several sequentially spaced slots and may have coil sides spaced less than 180° to improve the voltage waveform. When the rotor is driven by its motor, the rotating magnetic field of its poles generates voltages in the stator coils. These voltages are shown in Fig. 3.5 over one revolution of the rotor. They are seen to be a three-phase set, each phase displaced 120° from the other. The frequency of the voltages is proportional to the speed of the rotor. For a frequency of 60 cycles per second (Hz), the two-pole rotor must turn at 60 × 60 = 3600 r/min.

For a rotor with more than two poles, the speed to obtain 60-Hz voltage is 3600/p r/min, where p is the number of pole pairs. The rotor shown in Fig. 3.4b has four poles; that is, p = 2. The necessary speed to obtain 60-Hz voltage is 3600/2 = 1800 r/min. The stator coils for the four-pole generator are connected additively as shown in Fig. 3.4c.

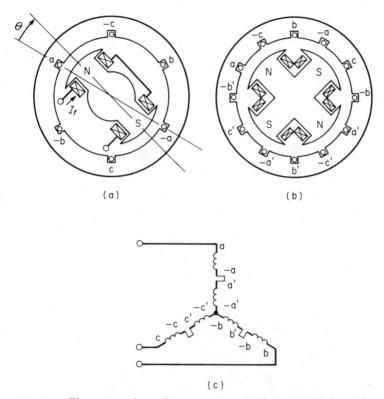

(a)   (b)

(c)

Figure 3.4 Elementary three-phase generator. (a) Two-pole; (b) four-pole; (c) Y connection of the windings [3].

Field excitation

The voltage is generated in the stator windings by the rotating magnetic field produced by the field poles on the rotor. The relationship between the stator voltage at no load and the rotor field current, termed the "excitation curve," is shown in Fig. 3.6. Initially, the voltage is linearly related to the field current. At rated voltage and higher, the magnetic material in the generator saturates and increas-

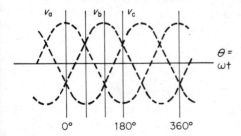

Figure 3.5 Voltage induced in the coils of Fig. 3.4a.

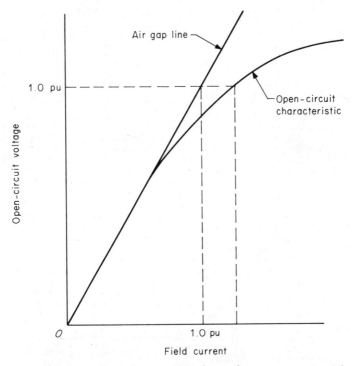

Figure 3.6   Open-circuit characteristic of a synchronous generator. 1.0 pu voltage represents the rated value. 1.0 pu field current represents the value for an unsaturated machine [3].

ingly higher field current is required to increase the stator voltage. Under load, the stator voltage drops below the no-load value for two reasons: (1) the resistance and leakage reactance of the stator windings and (2) the armature reaction of the stator currents, which tends to "demagnetize" the rotor poles. The stator winding is represented electrically by a voltage source of the no-load value and an impedance with components representing (1) and (2).

In an MG set, the ac power for the field current is obtained from the ac or dc source that drives the motor. The field power is rectified in the excitation system and controlled by the voltage regulator for constant stator voltage. The current is applied to the field windings through slip rings as shown in Fig. 5.6.

3.3   Synchronous/Synchronous MG Set

Functionally, a synchronous motor is built in the same way as a synchronous generator. To provide torque, the synchronous motor must operate in synchronism with the supply line, e.g., at a speed of 3600 r/min for a two-pole motor operating from a 60-Hz supply line. Consequently, it requires a means for starting and synchronizing with the supply line. Damper windings provided on the rotor field poles allow the machine to start as an induction motor; the machine synchronizes when direct current is applied to the rotor field windings. The synchronous generator also is equipped with damper windings to prevent electromechanical oscillation when two or more units are operated in parallel on a common bus.

A synchronous motor coupled to a synchronous generator forms the set. If the machines are direct-coupled, the ratio of the frequencies is that of the poles. For example, for the same pole number, the input and output frequencies are the same, but for a 4-pole motor and a 28-pole generator, the ratio is 7; a 60-Hz input yields 420-Hz output. If the machines are coupled by a gearbox or a set of pulleys and belt, the frequency ratio is the product of the pole ratio and the speed ratio. Any exact ratio can be obtained.

In a synchronous generator, the field current determines the stator terminal voltage; in a synchronous motor, it determines the power factor that the motor presents to the supply line. For each mechanical load on the motor, one value of field current will provide unity power factor. For field current greater than that value, the motor is overexcited; it delivers reactive power back to the supply line and appears as a leading power factor load. For field current less than that value, the motor is underexcited; it appears as a lagging power factor load. One feature of the synchronous motor is that it can compensate for the lagging power factor of other loads in the facility and improve the overall power factor or reduce the demand kVA on the utility.

One compact MG set commonly used to buffer computers contains two sets of stator windings of the same pole number on common stator laminations. The coils of the motor and generator are placed in alternate slots. One field structure serves both machines. The field current on the single winding is adjusted to regulate the terminal voltage of the generator. The power factor of the synchronous motor is not controllable; by design it is made to be about 0.9 at full load.

3.4 Induction/Synchronous Set

An induction motor, also termed an "asynchronous motor," operates by "slipping" with respect to the synchronous speed corresponding to the line frequency. The induction motor has a stator winding like that of a synchronous machine, but its rotor winding has no external connections and is instead short-circuited upon itself. The rotor winding consists of axial copper bars in the slots of the laminations that terminate at each end of the rotor in end rings. The bar and the end-ring

Figure 3.7   Cutaway view of a three-phase squirrel-cage induction motor with encapsulated windings. (*Courtesy The Lincoln Electric Co.*)

construction is termed a "squirrel cage." A squirrel cage can also be formed by die-casting aluminum bars and rings in place, as shown in the rotor of Fig. 3.7.

In operation, the slip of the rotor bars with respect to the magnetic field induces low-frequency currents in the bars which interact with the magnetic field to produce torque, in the manner of the field winding current in the synchronous motor rotor. The induction motor torque increases linearly with the slip in the operating range of the motor.

When an induction motor is coupled to a synchronous generator, the output frequency at no load is close to the ratio of the poles and any gears or pulleys used in the mechanical connection. As the generator is loaded, the induction motor speed slips; and the generator frequency also declines slightly from the no-load synchronous value. The full-load slip is about 1 percent for large induction motors and up to 5 percent for small ones. For example, a 1750 r/min induction motor operates at a slip of 50 r/min, or 2.78 percent, with respect to synchronous speed of 1800 r/min.

The major use of induction/synchronous MG sets is to supply 415-Hz power to computers. A typical set consists of a 2-pole 3600 r/min induction motor driving a 14-pole synchronous generator. The frequency drops from 420 Hz at no load to 415 Hz at full load. These sets are designed with rotating-rectifier exciters for the generator and "soft starters" for the motor to make the starting current commensurate with the running current. They also have ride-through capability when voltage dips occur in the 60-Hz supply line.

Induction/synchronous MG sets operate easily with the generator terminals paralleled on a common output bus. The speeds are held identical by the self-synchronizing torques between the generators. When a set is started in preparation for synchronizing it with an already loaded set or group, its speed is higher than that of the loaded set or sets. The set must first be loaded with an off-line (dummy) load to bring its speed down to that of the loaded sets. When the frequency and phase of the generator voltage are the same as those of the operating bus, the generator breaker can be closed and the off-line load switched off; the set will then assume its part of the total electrical load. The dc field currents of the generators will be adjusted automatically by their voltage regulators to maintain correct bus voltage and to ensure that the generators divide the reactive-power load.

3.5 Ride-Through Capability

Most supply line disturbances last less than a second. The MG set buffers the load from such disturbances by using the stored kinetic energy in its rotating structure to supply the electrical load. The capability of one type of set is shown in Fig. 3.8. The voltage envelope is typical of the acceptable input to the computer. In this case, the com-

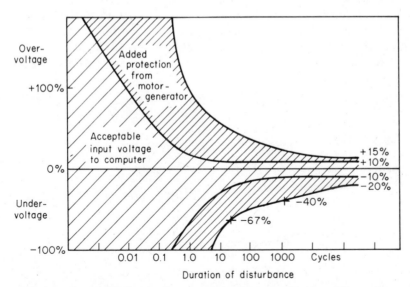

Figure 3.8 Acceptable input voltage to a computer vs. duration of line voltage disturbance. Added protection from the motor-generator set is shown as a ride-through capability.

puter can tolerate zero line voltage for up to one-half cycle (8 ms). The dark region shows the extra line voltage tolerance that the MG set provides.

The MG set envelope shows the limits of +15 and −20 percent for the line voltage to the drive motor for which the generator of the set will deliver acceptable voltage to the computer. During a line voltage dip on all three phases to zero volts for 100 ms, the generator will deliver acceptable voltage to the computer. During a 67 percent dip for 0.5 s or a 40 percent dip for 16 s, the voltage will be acceptable. During a 15 percent dip, the set will deliver acceptable voltage continuously.

3.6   Uninterruptible MG Set

The MG set has no long-time energy storage capability. To provide uninterruptible operation of ac-to-ac MG sets in the face of utility power failure, one of the three following approaches can be used:

1. Supply the MG set from the 60-Hz output bus of a battery-supported static UPS. The MG set can receive power initially from the batteries and, later, from backup engine-generators. Such an installation depends upon whether the UPS has spare power available and whether the MG set is designed to "soft start" so as not to dip the UPS output bus voltage. That is common with 60- to 415-Hz MG sets.

2. Provide a standby forced-commutated inverter, as shown in Fig. 3.9, to supply power to the MG set from the battery when the normal line fails. Because the MG set can satisfy the load for about 100 ms without input power, electromagnetic contactors such as MI and ML, rather than static switches, can be used for the transfer from the normal line to the inverter.

3. Provide a load-commutated inverter, as shown in Fig. 3.10, for a

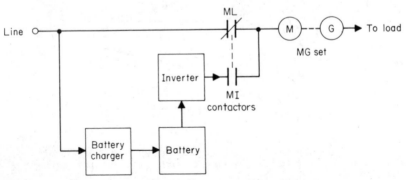

Figure 3.9   MG set with standby forced-commutated inverter.

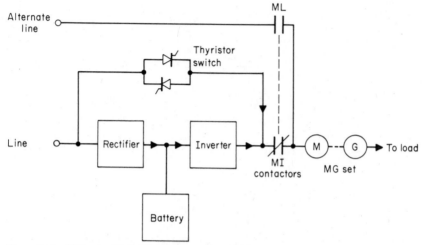

Figure 3.10   MG set with load-commutated inverter.

synchronous/synchronous MG set. A relatively simple inverter can be used to drive the synchronous motor; the internally generated voltages in the motor will commutate the inverter thyristors at the proper instants. In the commercial version, the inverter idles at 5 percent in normal operation; the remainder of the power is supplied through the thyristor switch. An alternate line is provided for driving the motor when the inverter and thyristor switch are out of service.

Additional approaches include: (1) a dc-motor-driven synchronous generator, for which drive power is obtained from a battery charger rectifier and battery, (2) an ac-ac MG set with a flywheel and a diesel engine, as shown in Fig. 3.11, and (3) an ac-dc-ac MG set, in which the

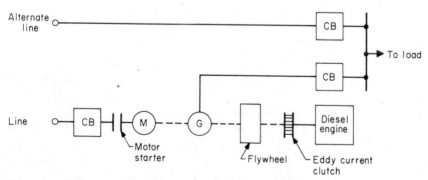

Figure 3.11   MG set with diesel engine support.

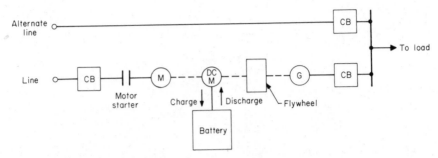

Figure 3.12 MG set with dc motor and battery support.

dc machine acts as both the battery charger and the standby drive motor, as shown in Fig. 3.12. An actual unit is shown in Fig. 3.2.

Example 3.1 *Motor rating.* Find the minimum horsepower rating of a standard induction motor to drive a synchronous generator with an output of 100 kVA, 0.8 PF, 80 kW and an efficiency of 0.92 at full load.

solution The calculated motor horsepower at generator full load is

$$\text{Horsepower} = \frac{80 \text{ kW} \times 0.746 \text{ hp/kW}}{0.92} = 64.9 \text{ hp}$$

The minimum size standard induction motor is 75 hp.

Example 3.2 *Battery bank rating.* Find the ampere-hours required from a battery bank to drive a dc-motor/synchronous-generator set at full load for 5 min. The data are as follows: generator, 200 kVA, 0.8 PF, 0.95 efficiency; motor, 125 hp, 240 V, 0.92 efficiency.

solution The power required from the battery is found by

$$\text{Power} = \frac{200 \text{ kVA} \times 0.8}{0.92 \times 0.95} = 183 \text{ kW}$$

The ampere-hour requirement at 240 V dc for 5 min is found by

$$\text{Ah} = \frac{183 \text{ kW} \times 10^3 \times 5 \text{ min}}{60 \text{ min/h} \times 240 \text{ V}} = 63.5 \text{ Ah}$$

3.7 Summary

Motor-generator sets utilize either induction, synchronous, or dc motors to drive ac generators for supplying power to critical loads. The sets buffer the loads from short-time transients in the normal utility supply. The kinetic energy of the set can provide power when the utility power fails for up to 100 ms. The time can be extended up to about one second by adding a flywheel, up to 5 to 20 min by adding a dc motor and a battery bank, and up to several hours by adding a diesel engine.

REFERENCES

1. ANSI/NEMA MG1-1978, "American National Standard for Motors and Generators."
2. ANSI/NFPA 70-1987, "National Electrical Code®."*
3. A. E. Fitzgerald, C. Kingsley, and S. Umans, *Electric Machinery,* 4th Ed., McGraw-Hill, New York, 1983.

---

*National Electrical Code® and NEC® are registered trademarks of the National Fire Protection Association, Inc., Quincy, Mass.

Chapter

4

Transfer Switches

A transfer switch is a device that transfers a common electrical load from a normal supply to an alternate supply in the event of failure of the normal supply and returns the electrical load to the normal supply when the normal supply is restored. It may be operated automatically from relays that sense the normal supply voltage and/or frequency or manually from a local or remote control point.

The transfer switch is either a self-contained unit with two sets of contacts or an assembly from two electrically or mechanically interlocked contactors or circuit breakers. The automatic transfer switch (ATS) includes the relays and other controls that determine the operations of the switch. A transfer switch can also be built with a bypass switch which enables the transfer switch to be electrically isolated, jacked out, or completely removed for maintenance and testing. Transfer switches rated for secondary voltages up to 600 V ac are available in capacity from 100 to 4000 A. They are also rated for short-circuit withstand and closing ratings in accordance with ANSI/UL-1008 [1]. Transfer switches for primary voltages, e.g., 4.16, 6.9, or 13.2 kV, are assembled from standard circuit breakers for those voltages and are electrically interlocked.

4.1 Codes and Standards

The requirements for transfer switches are given in the following documents:

1. Underwriters Laboratory, ANSI/UL 1008, "Automatic Transfer Switches" [1]

2. ANSI/NFPA 70-1987, "National Electrical Code®,"* Arts. 517, 700, 701, 702 [2]

---

*National Electrical Code® and NEC® are registered trademarks of the National Fire Protection Association, Inc., Quincy, Mass.

3. ANSI/NFPA 110-1985, "Emergency and Standby Power Systems," Chap. 4, "Electrical-Switching and Protection" [3]

Considerable information on construction, ratings, and application is provided in manufacturers' bulletins.

4.2 Types of Transfer Switches

Transfer switches are classified as automatic (self-acting) or nonautomatic (direct manpower or electrical remote manual control). Within those classifications they can be described more specifically by the following:

1. *Single operator.* Automatic transfer switch, which is a double-throw switch actuated by a single electrical operator, e.g., motor, with a fixed contact transition time up to 0.5 s. The transfer switch may also have a manual operator that transfers the contacts in the same time in lieu of the electrical operator. The switch is available with two, three, or four poles for one-phase, three-phase, and three-phase with neutral circuits.

2. *Dual operator.* Automatic transfer switch, which is a double-throw switch actuated by two electrical operators. The contact transition time is adjustable; typically it is up to 300 s. The switch is designed primarily for transferring the electric power supply to large motors and transformers. The transfer switch may also have a manual operator that transfers the contacts in the adjustable transition time in lieu of the electrical operator.

3. *Manual control.* Non-automatic-transfer switch, which is a double-throw switch actuated by a single electrical operator with a fixed transition time. The switch is operated manually by push buttons, locally or remotely, rather than automatically by relays. It may also have a manual operator (handle) that transfers the contacts in lieu of the electrical operator.

4. *Bypass/isolation.* The automatic transfer switch is combined with a separate bypass/isolation switch in an integrated unit. The bypass portion provides the means for connecting either the normal or the alternate source directly to the load, in case the transfer switch is inoperative. The isolating portion provides the means for deenergizing and isolating the transfer switch for maintenance, testing, or repair.

5. *High voltage.* Primary-voltage-level circuit breakers, combined to switch one of two electric power sources to the load. The two breakers can feed into a common bus or into two buses using a third bus-tie

breaker. The operation and interlocking of the circuit breakers are done by relays.

4.3 Construction

A single-operator automatic transfer switch is shown in Fig. 4.1. The parts of the switch are the following:

A   Terminals for connection to the load
B   Terminals for connection to the normal source

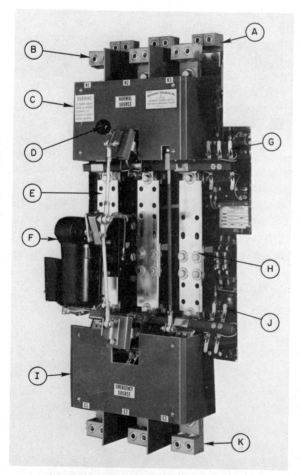

Figure 4.1 Single-operator automatic transfer switch. (*Courtesy Russelectric, Inc.*)

C   Normal switch contacts

D   Manual operator

E   Mechanical interlock between the upper normal switch and the lower emergency switch

F   Electric motor operator

G   Auxiliary contacts for normal switch

H   Load buses

I   Emergency switch contacts

J   Auxiliary contacts for auxiliary switch

K   Terminals for connection to the alternate source

During a transfer, the contact mechanism is securely locked in position during the first half of the operation. When the operator reaches the over-center position, the preloaded springs snap open the closed contacts and snap closed the open contacts with a momentary break in between. The quick break ensures full arc interruption in less than one-half cycle.

The contact mechanism and arc chutes of the switch shown in Fig. 4.1 are shown in Fig. 4.2. The contact assembly utilizes segmented silver-tungsten main contacts and separate silver-tungsten arcing contacts. The arc chutes consist of parallel steel plates partially surrounding the contacts and enclosed by a ceramic insulator. When the contacts open, the arc is drawn into the plates, which splits it into a series of smaller arcs, thereby cools it, and finally extinguishes it.

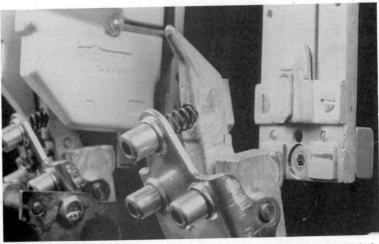

Figure 4.2 Contact mechanism and arc chutes of automatic transfer switch. (*Courtesy Russelectric, Inc.*)

A dual operator transfer switch incorporates two operators like the electric motor F in Fig. 4.1. The construction of the transfer switch with bypass/isolation features will be described in Sec. 4.4.

4.4 Operation

All transfer switches operate to shift or transfer electrical load between a normal and an alternate source. Several typical applications will be described.

Utility feeders

A normal and an alternate utility feeder are commonly used for facilities that may already utilize engine-generators and UPSs as alternate sources. The alternate utility feeder can carry all of the facility load, whereas the engine generator and UPS are designed to carry only the emergency load.

A transfer switch arrangement whereby the switching is done on the secondary side is shown in Fig. 2.4. The control relays will initiate the transfer from the normal feeder when the voltage on any phase drops below a preset value, typically 80 percent, for an adjustable time delay of 0.5 to 3 s, to exclude transfer for momentary dips. The relays will permit the transfer only if the voltage of the alternate feeder is within limits. After the transfer is completed, the return transfer to the normal feeder can be initiated manually or automatically. The acceptable quality of the normal feeder voltage is preset, typically above 90 percent on all phases for a duration of 25 min.

Standby generator

An emergency system utilizing three generators and three transfer switches is shown in Fig. 4.3. The generator system is designed to automatically supply emergency power during a utility power outage. Emergency power is derived from three engine-generator sets. In the event of a utility power failure, the engines will automatically start, parallel, and supply emergency power to the associated loads.

When the engine-generators and the master control switches are in their automatic positions, the emergency generating system is placed on standby. Upon receipt of an engine start signal from the automatic transfer switch system, all engine-generators will start, and the first to reach 90 percent of rated voltage and frequency is connected to the emergency bus through its generator circuit breaker within 7 s of the start signal.

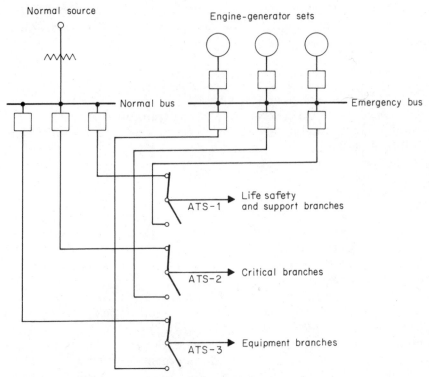

Figure 4.3 Standby generators; emergency system for a hospital.

At this time, pilot contacts will operate simultaneously to permit the connection of load to the bus through the life safety and critical automatic transfer switches, ATS-1 and ATS-2. Within 10 s of the engine start signal, the remaining generators will automatically synchronize and be paralleled to the emergency bus. When the second generator is on line, pilot contacts will allow the equipment load automatic transfer switch, ATS-3, to operate and transfer its load to the bus. A complete alarm and failure system is provided to automatically shut down an engine-generator.

In the event of an engine-generator failure, the nonessential equipment load circuits are automatically shed by the transfer switch, which transfers to an off position until an acceptable source is available; at which time it will transfer to that source. This assures continuous power to the critical and life safety loads.

When the utility supply voltage returns, the transfer switch system will transfer all loads back to the utility source. The generator circuit breakers will simultaneously open, and the engines will continue to run through a cool-down period and then shut down. All controls will now automatically reset and be in readiness for the next operation.

Bypass/isolation switch

A schematic diagram of the bypass/isolation switch adjunct to the automatic transfer switch is shown in Fig. 4.4; the physical arrangement is shown in Fig. 4.5. The bypass feature is used when the transfer switch is inoperative. The isolation feature is used when the transfer switch must be disconnected from the source and load circuits, as for maintenance.

With the bypass handle and the isolating handle of the assembly in Fig. 4.4 in the normal positions, the automatic transfer switch operates normally. When contacts a are closed, the load is connected directly to the load terminal of the transfer switch. Closing contacts d connects the normal source; closing contacts e connects the alternate source. When the bypass handle is placed in the N (normal) position, contacts a open and contacts b close to connect the normal source di-

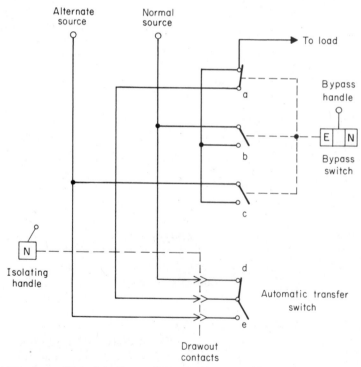

Figure 4.4 Schematic diagram of bypass and isolation switch.

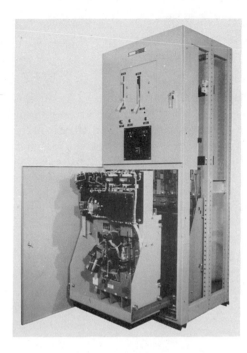

Figure 4.5 Transfer switch with provision for bypass and isolation. (*Courtesy Russelectric, Inc.*)

rectly to the load. When the bypass handle is placed in the E (emergency) position, contacts *a* open and contacts *c* close to connect the alternate source directly to the load. The route through the transfer switch is bypassed.

When the bypass handle is in either the N or E position, the isolating handle can be operated to mechanically open the drawout contacts of the transfer switch. The transfer switch can be removed without disturbing the supply to the load.

Primary transfer switch

Transfer switches for operation at primary voltage are assembled from conventional circuit breakers and controlled by separate relays. The three primary circuit breakers shown in Fig. 4.6 are electrically interlocked and can be operated in the following modes:

1. *Load can be taken from both sources.* For normal operation, breakers *a* and *b* are closed and *c* is open. If either source fails, its breaker opens and the tie breaker *c* closes, which places all of the load on one source. If the source cannot carry the total load, nonessential load must be shed before the transfer takes place.

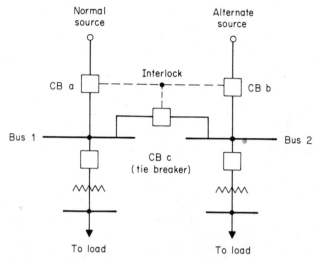

Figure 4.6 Primary transfer switch arrangement.

2. *Load can be taken from only one source.* Normally, the normal source breaker a is closed, the alternate source breaker b is open, and the tie breaker c is closed. When the normal source fails, its breaker a opens and the alternate source breaker b closes.

4.5 Controls

The automatic transfer switch must be electrically operated and mechanically held. ANSI/NFPA-110 [3] requires a minimum of the following controls for a transfer switch in a system with an engine-generator alternate source (EPS):

1. *Source monitoring.* (*a*) Undervoltage-sensing devices shall be provided to monitor all ungrounded lines of the normal source of power. When the voltage on any phase falls below the minimum operating voltage of any load to be served, the transfer switch shall automatically initiate engine start and the process of transfer to the EPS. When the voltage on all phases of the normal source returns to within acceptable limits, the load shall be retransferred to the normal source. (*b*) Both voltage- and frequency-sensing equipment shall be provided to monitor one ungrounded line of the EPS power. Transfer to the EPS shall be inhibited until there is adequate voltage and frequency to handle loads to be served.

2. *Interlocking.* Reliable mechanical interlocking or an approved alternate method shall prevent the inadvertent interconnection of the

normal power supply and the EPS or of any two separate sources of power.

3. *Manual operation.* Means shall be provided for safe manual nonelectric transfer in the event the transfer switch should malfunction.

4. *Time delay on starting of EPS.* A time delay device shall be provided to delay starting of the EPS. The timer is intended to prevent nuisance starting of the EPS with subsequent load transfer in the event of harmless momentary power dips and interruptions of the normal source.

5. *Time delay on transfer to EPS.* An adjustable time delay device shall be provided to delay transfer to the EPS when the transfer switch is installed for Level 1 use and loads must be sequenced to avoid excessive voltage drop. The time delay shall commence when proper EPS voltage and frequency are achieved.

6. *Time delay on retransfer to normal source.* An adjustable time delay device with automatic bypass shall be provided to delay retransfer from the EPS to the normal source of power. The timer is intended to permit the normal source to stabilize before retransfer of the load. The time delay shall be automatically bypassed if the EPS fails.

7. *Time delay on engine shutdown.* A time delay of 5 min minimum shall be provided for unloaded running of the EPS prior to shutdown. This delay provides additional engine cooldown. This time delay shall not be required on small (15-kW or less) air-cooled prime movers.

8. *Engine generator exercising timer.* A program timing device shall be provided to exercise the EPS. When load is switched, immediate return to normal power shall be automatically accomplished in the event of EPS failure.

9. *Test switch.* A test switch that will simulate failure of the normal power source shall be provided on each automatic transfer switch. ATS shall transfer the load to EPS.

10. *Indication of switch position.* Two pilot lights with identification nameplates, or other approved position indicators, shall be provided to indicate the transfer switch position.

The control diagram for an automatic switch in the normal position is shown in Fig. 4.7 [4]. When the normal source fails or the voltage falls to a predetermined point, e.g., 70 percent, the phase relays B1, B2, B3 are deenergized. The main control relay CR drops out, which

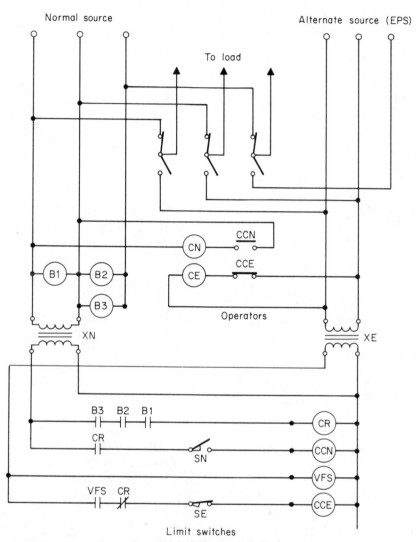

Normal source

Alternate source (EPS)

To load

CCN

CN

CCE

CE

B1   B2

B3

Operators

XN

XE

B3 B2 B1

CR

CR

CCN

SN

VFS

VFS CR

CCE

SE

Limit switches

Figure 4.7 Control schematic diagram for automatic transfer switch.

closes a contact in the circuit to the emergency control relay CCE and provides a signal to start the alternate source, e.g., the engine-generator set. When the alternate source reaches 90 percent of rated voltage and frequency, control relay VFS closes. With the limit switch SE closed, the emergency control relay CCE is energized, which energizes the solenoid operator CE. The main contacts disconnect the load from the normal source and connect it to the alternate source. When the solenoid completes its stroke, the limit switch SE opens, which deenergizes relay CCE and the solenoid operator CE. The transfer switch is now mechanically locked into the alternate (emergency) position.

4.6 Static Transfer Switch

Automatic electromechanical transfer switches, particularly when combined with engine-generator sets, fulfill most of the requirements for emergency/standby systems [5]. However, there are several applications in which the electromechanical switch sense-and-transfer time of 4 to 20 cycles is too long.

The first, obviously, is the power supply for data processing and communications equipment that cannot tolerate a line-voltage interruption of more than one-half cycle. The second is to high-pressure mercury arc fixtures, which require up to several minutes to cool and restart after a momentary loss of line voltage. The third is to electric motors for air-conditioning systems and MG sets that may be required to maintain the facility. If voltage is reapplied to these motors in the period up to 20 cycles when their air-gap magnetic fields have not decayed, but their voltages are out of phase, a surge in motor current that may damage the motor and trip the feeder breaker will occur.

One way to remedy the problems of the electromechanical switch for the above applications is to employ a solid-state transfer switch that can sense and transfer in the order of one-quarter cycle. With a two-source system, such as with a double-ended substation with two feeders, the solid-state switch can transfer rapidly enough to maintain the critical load.

One pole of a static ac switch consists of two back-to-back thyristors plus their control and firing circuits. A thyristor is a solid-state device which conducts current in one direction following the application of a firing signal to its gate terminal. In an ac circuit, the thyristor conducts current until the end of the half cycle; when the current attempts to reverse, the thyristor turns off. Two thyristors can be arranged to conduct both halves of the current wave and thus act as an ac switch. Thyristors are available in ratings up to 3500 V and 1800 A. Six thyristors are required for a three-phase switch.

An arrangement of two sets of thyristors to form a transfer switch is shown in Fig. 4.8 [6]. In this case the transfer switch supplies an ac line-voltage regulator from either a normal or an alternate source. The static switch can transfer the sources in as little as one-quarter cycle. The combination shown in Fig. 4.8 acts as an uninterruptible power source for computers and data processing equipment that are sensitive to loss of supply voltage for one-half cycle or more. A manual bypass switch and isolating circuit breakers are provided for operation

Normal   Alternate
source   source

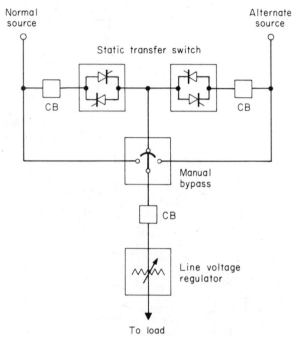

Figure 4.8 Static transfer switch controlling dual source to
line voltage regulator [6].

with a failed switch and for its repair. Static transfer switches are
available for 100 to 2500 A, 120 to 600 V, one-phase and three-phase.

Figure 4.9 shows a data processing center which is supplied from a
double-ended substation with feeders A and B [5]. Two static transfer
switches are used. One switch feeds the computer, computer peripherals, and emergency security load. The other switch feeds the essential
air-conditioning and emergency lights, including high-pressure
mercury-vapor fixtures. If a voltage deviation or failure occurs at either switch, the load is transferred to the other substation bus. An
engine-generator set can be provided to serve the loads if both buses of
the substation lose power.

Example 4.1   An emergency system consisting of an engine-generator set and a
transfer switch is shown in Fig. 4.10. Determine the following: (1) rating of the
generator to accommodate a 20 percent growth in emergency load, (2) rating of
the transfer switch, and (3) whether current-limiting fuses are required in the
circuits to the transfer switch.

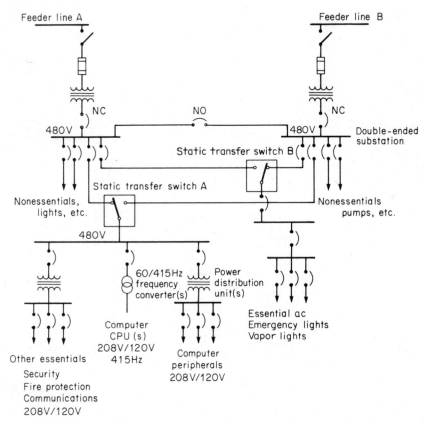

Figure 4.9   Data processing center supplied from double-ended substation and two static transfer switches [5].

solution   *Generator rating.* The future emergency load is

$$600 \text{ kW} \times 1.2 = 720 \text{ kW}$$
$$750 \text{ kVA} \times 1.2 = 900 \text{ kVA}$$

Generators are available in ratings of 875 kVA/700 kW, 1000 kVA/800 kW, and 1125 kVA/900 kW. Select the generator rating at 1000 kVA/800 kW. The rated current is 1203 A at 480 V.

Transfer switch rating. Based on future load, the transfer switch rating is 900 kVA, 1083 A; based on the generator, it is 1000 kVA, 1203 A. Transfer switches are available in ratings of 1000, 1200, and 1600 A. Select the transfer switch rating of 1600 A.

Current-limiting fuse requirement. The 1600-A transfer switch has a withstand and closing rating into a fault when coordinated with a circuit breaker of 85,000 A and with a current-limiting fuse of 200,000 A.

For the fault on the load side of the transfer switch and for the switch in the normal position, the fault current delivered by the transformer is given by

$$I \text{ (fault)} = \frac{2500 \text{ kVA} \times 10^3}{1.732 \times 480 \text{ V} \times 0.05 \text{ pu}} = 60{,}142 \text{ A}$$

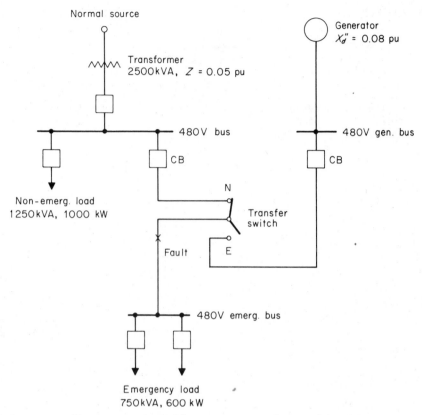

Normal source

Generator
$X_d'' = 0.08$ pu

Transformer
2500kVA, $Z = 0.05$ pu

480V bus

480V gen. bus

CB

CB

Non-emerg. load
1250kVA, 1000 kW

N

Transfer
switch

Fault

E

480V emerg. bus

Emergency load
750kVA, 600 kW

Figure 4.10   Emergency system with one generator and a transfer switch.

and with the switch in the emergency position, the fault current delivered by the generator is given by

$$I \text{ (fault)} = \frac{1000 \text{ kVA} \times 10^3}{1.732 \times 480 \text{ V} \times 0.08 \text{ pu}} = 15,036 \text{ A}$$

Current-limiting fuses are not required. The switch and circuit breakers can withstand and clear 85,000 A. The respective fault currents are 60,142 and 15,036 A.

4.7 Summary

The transfer switch is the heart of most emergency systems. It enables critical loads to be supplied selectively from the normal utility line or from alternate sources. Commercial transfer switches are supplied with the motor operators and the relays to sense the conditions under which they must operate.

REFERENCES

1. ANSI/UL 1008-1983, "Safety Standard for Automatic Transfer Switches."
2. ANSI/NFPA 70-1987, "National Electrical Code®."
3. ANSI/NFPA 110-1985, "Emergency and Standby Power Systems."
4. "Zenith ZTS Transfer Switches," Zenith Controls Inc., Chicago, Ill., 1985.
5. D. C. Griffith, "When and Where for the Static Switch," *Conf. Proc. Intelec '86,* 1986, pp. 451–456.
6. D. C. Griffith, "Trends in A.C. Power Conditioning for Telecommunications," *Conf. Proc. Intelec '85,* 1985, pp. 87–91.

Engine-Generator Sets

Engine-generator sets are an electric energy source alternative to the normal utility supply. The sets serve the following functions:

1. Continuously operating independent electric energy sources, particularly at sites having no convenient utility service.

2. Demand reduction (peak-sharing) at installations where the sets have other functions or the economics of the utility demand charge favors the operation of the sets.

3. Cogeneration, where the engine also supplies heat or hot water to the facility. Part of the electric energy produced may be sold to the utility.

4. Standby for emergency systems, where the sets are started and pick up emergency load when the normal utility supply fails.

NFPA 110 [1] defines sets using rotating equipment as emergency power supplies (EPS) in emergency power supply systems (EPPS). This includes rotating generators driven by engines, gas turbines, and steam turbines. The bulk of the EPS equipment is engine-generator sets.

5.1 Standards

The requirements for engine-generator sets are given in the following documents:

1. ANSI/NFPA 110-1985, "Emergency and Standby Power Systems," Chap. 3, "Power Supply: Energy Source, Converters and Accessories" [1]

Figure 5.1 Gasoline engine-generator set rated 15 kW. (*Courtesy Kohler Co.*)

2. ANSI/NFPA 70-1987, "National Electrical Code®,"* Art. 445 [2]

3. ANSI/NFPA 37-1984, "Installation and Use of Stationary Combustion Engines and Gas Turbines" [3]

5.2 Types

Emergency power supplies (EPS) using rotating equipment (as compared to battery-powered) can be classified by energy source (fuel) and by prime mover as follows:

1. Liquid petroleum products at atmospheric pressure

2. Liquified petroleum gas

3. Natural or synthetic gas

The sets are classified by prime mover as follows:

1. *Gasoline engine.* Sets are available from several hundred watts to about 100 kW. A typical set is shown in Fig. 5.1. Smaller sets use two- and four-cycle high-speed lightweight engines. Larger sets use multicylinder engines built for automobiles and trucks. For direct-connected 60-Hz generators, the engines must run at $3600p$ r/min, where p is the number of pole pairs, e.g., for two poles, $p = 1$, speed = 3600 r/min. Gasoline engines can run on all three types of fuel listed above.

---

*National Electrical Code® and NEC® are registered trademarks of the National Fire Protection Association, Inc., Quincy, Mass.

2. *Diesel engine.* Sets are available from several hundred to over 10,000 kW. A typical set is shown in Fig. 5.2. Engine speeds range from 1800 r/min for the lower ratings to 600 r/min for the higher ratings. Diesel engine sets are usually designed to run for long periods of time, but they are heavier and costlier than gasoline engine sets. The sets require about 10 s to start, reach rated speed, and deliver power. Diesel engine sets by far dominate the types used for emergency/standby systems.

3. *Gas turbine.* Sets are available up to 10,000 kW. A typical set is shown in Fig. 5.3. The speed of the turbine is stepped down in a gearbox to 1800 or 3600 r/min to drive the 60-Hz generator. These sets are compact and lightweight; they are suitable for mounting in restricted space and on the roofs of buildings. The gas turbines are modified aircraft auxiliary power and small propulsion power turbines. The sets require about 120 s to start, reach rated speed, and deliver power.

5.3   Generators

Engine-driven generators are rated by speed, frequency, voltage, and power. They fall into two general categories: (1) high-speed 3600 r/min

Figure 5.2   Diesel engine-generator set: 175 kW, 218.75 kVA, 480Y/277 V, 60 Hz. (*Courtesy Katolite Corp.*)

Figure 5.3 Gas turbine–generator set rated 3300 kW. (*Courtesy Ruston Gas Turbines, Inc.*)

2-pole or 1800 r/min 4-pole machines of moderate diameter with rotor supported on two bearings; (2) low-speed 720 r/min 10-pole or 600 r/min 12-pole machines of large diameter compared to axial length with rotor supported on one generator bearing and one engine bearing.

Operation

The construction and operation of a synchronous generator are described in Sec. 3.2.

Exciters

The exciter delivers field current to the generator in order to magnetize the field poles and thereby control the generated voltage of the stator windings. Unlike the MG set, which can obtain its field power from the same source as the drive motor, the engine-driven generator must provide its own field power by means of an auxiliary generator, called an exciter, or by a "bootstrap" method using part of its own output.

The basic excitation system is shown in Fig. 5.4. The field current is supplied to the generator through brushes in contact with slip rings. The field power is developed by a small rotating generator called an exciter. In the figure an ac exciter with rotating permanent magnet field poles is shown. The output is rectified, controlled by the voltage

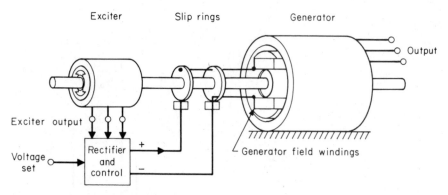

Figure 5.4   Slip ring rotating exciter.

regulator, and applied to the slip ring brushes. The exciter in Fig. 5.4 can also be a small dc generator.

A second type of excitation system, called a brushless exciter, is shown in Fig. 5.5. The brushes and slip rings are eliminated. The exciter is an ac generator with stationary field poles; the ac voltage generated in the rotating windings is rectified by diodes mounted on the rotating structure. The resultant dc voltage produced by the rotating rectifiers is applied directly to the field windings of the generator. The voltage regulator controls the exciter field current to obtain the desired generator terminal voltage.

A third type of excitation system, called a static exciter, eliminates the rotating exciter but retains the slip rings and brushes, as shown in Fig. 5.6. The field power is derived from the output of the generator. Part of the power is supplied by a current transformer, current I_c, and part by a potential (voltage) transformer, current I_p. By combining the two components of current by using the inductor for phase shift, the static exciter provides slightly more than the required field current I_f for each load condition.

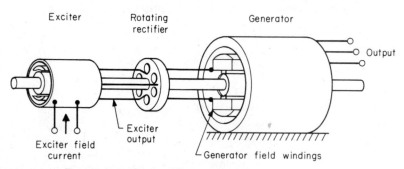

Figure 5.5   Rotating-rectifier brushless exciter.

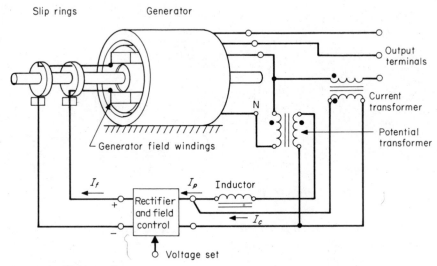

Figure 5.6   Static exciter.

The voltage control system then diverts the excess field current to regulate the generator voltage. The static exciter provides very fast recovery of the generator voltage after an initial dip by suddenly applied loads, such as motor starting. A dc source, such as the engine-starting battery, may be required to "flash" the exciter.

5.4   Generator Voltage Waveform

The ideal voltage waveform is described as a sine wave with respect to time at the rated frequency of the generator, e.g., 60 Hz. The actual voltage waveform differs from the ideal because of the design of the generator and the effect of the load current. The difference is described and specified in two ways: (1) by the harmonic voltage components and (2) by the deviation from an ideal sine wave. The harmonic voltage components, typically the odd harmonics, are expressed as the rms harmonic voltage V_h by

$$V_h = (V_3^2 + V_5^2 + V_7^2 + \ldots)^{1/2}$$

where $V_3, V_5, V_7, \ldots$ are the rms values of each harmonic voltage component. The value of V_h is usually expressed as a percent of the rms fundamental voltage V_1 by

$$V_h \text{ (percent rms)} = \frac{V_h}{V_1} \times 100$$

The deviation from an ideal sine wave is determined by drawing an ideal sine wave through the actual voltage waveform and measuring the maximum deviation ΔV in volts from the ideal sine wave. The maximum deviation is expressed as a percentage of the peak of the ideal sine wave V_p by

$$\Delta V \text{ (percent deviation)} = \frac{\Delta V}{V_p} \times 100$$

One or both of the requirements can be specified for the line-to-neutral and/or line-to-line voltage and for the no-load and/or rated load conditions. Typically, 5 percent is specified for the V_h and ΔV.

Generator design

With the generator unloaded, the waveform of the stator voltages is determined by (1) the shape of the field pole, (2) the modulation of the magnetic field by the stator teeth, and (3) the attenuating effect on the harmonics of the stator winding design.

The field pole is so shaped that the amplitude of the magnetic field in the air gap over one pole pitch is distributed as a sine wave from 0° to 180°. That sine wave of magnetic field will produce a sine wave of voltage in each stator conductor and at the terminals of the generator. However, because the inner surface of the stator magnetic laminations is not smooth, but instead consists of teeth and slots, the magnetic field is modulated and may produce high-frequency tooth harmonics (ripple) in the stator voltage.

To compensate for distortion of the magnetic field in the air gap of the generator, particularly when the generator is loaded, the individual coils for each phase of the generator are so distributed in the slots that the sum of the induced harmonic voltages is either zero or less than the sum of the fundamental voltages in the coils that make up each phase. In addition, either the stator slots or the poles are skewed along their longitudinal axis by one stator slot pitch. The effect is to cancel the effect of the teeth on the magnetic field and to cancel the tooth harmonic voltages. See Ref. 4 for additional information.

Load current

When a generator is loaded, the load current affects the voltage waveform by (1) voltage drop, (2) armature reaction, (3) magnetic saturation, (4) load-current harmonics.

The voltage drop in a generator has three components. The first component is the drop in the resistance of the windings. The second is the drop in the leakage reactance of the windings. The third is the result of the demagnetizing effect on the field poles of the stator current. The voltage drop effect merely reduces the amplitude of the generator voltage. However, the third component, also called the armature reaction, modifies the magnetic field in the generator so that the induced voltage in the stator windings is distorted from the sine wave that it could have under no-load conditions.

The armature reaction also forces the voltage regulator for the generator to increase the field current to maintain the terminal voltage constant. The increased field current, which matches the stator current, can cause certain parts of the rotor field poles and the stator teeth to saturate magnetically. The effect is further distortion of the magnetic field and the induced voltage at heavy generator loads.

Independently of the generator design, the waveform of the load current can become distorted from a sine wave by "nonlinear" loads supplied by the generator. Typical nonlinear loads include power transformers operated near magnetic saturation, switching power supplies for computers, fluorescent and mercury arc lamp ballasts, and diode or thyristor rectifiers. A nonlinear load is one that draws a nonsinusoidal current from a sinusoidal applied voltage. The distorted nonsinusoidal load current, delivered through the internal impedance of the generator, distorts the terminal voltage of the generator for all loads supplied by the generator.

Voltage regulation

The terminal voltage of the generator must be regulated by means of the field current in the face of load change and temperature change of resistance of the field windings. The voltage regulator performs this function. As shown in Fig. 5.7, the regulator is a feedback-control system. The three-phase ac generator voltage is stepped down in transformers, rectified, and filtered as the feedback signal. It is compared with an adjustable reference voltage. The error difference is amplified and used to drive the exciter, which in turn supplies the generator field current. The error voltage is typically less than 1 percent of the

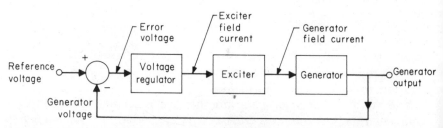

Figure 5.7 Block diagram of generator voltage regulator.

generator voltage at full load when it must drive the exciter to supply the maximum generator field current.

5.5 Generator Loading

In selecting ac generators for emergency/standby systems, the behavior of both the terminal voltage and the field current under steady-state load and under transient load may become critical. The terminal voltage may run out of an acceptable range for sensitive loads; the field current may run out of the range that can be delivered by the excitation system.

Generator impedance

The electrical representation of a generator for all loading conditions can be relatively complex [4]. An equivalent circuit for one phase of an unsaturated round-rotor machine (no projecting field poles) is shown in Fig. 5.8. The internal impedance of the generator consists of the stator winding resistance r_a, the stator winding leakage reactance x_l, and the magnetizing reactance x_m. The voltage V_t is the terminal voltage (line-to-neutral); the voltage E_a is the air gap voltage induced by the magnetic field; the voltage E_f is the excitation voltage proportional to the field current. The synchronous reactance is defined as $x_s = x_m + x_l$. The equivalent circuit of Fig. 5.8 can be used to calculate the effect of steady-state loads on terminal voltage and field current. For a generator with projecting field poles on the rotor, the representation must take into account that the magnetizing reactance x_m is higher along the axis of the poles than it is between the poles. The synchronous reactance x_s is segregated into a direct-axis reactance x_d and a quadrature-axis component x_q, and the calculation is carried out on each axis.

Under transient loading, as for short circuits, load switching, and motor starting, the generator is represented by the subtransient reactance x''_d for the initial part of the transient and by the transient re-

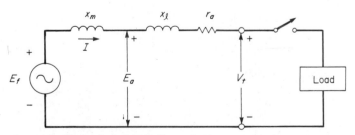

Figure 5.8   Equivalent circuit of round-rotor ac generator.

actance x'_d for the part after a few cycles. All of the generator impedances are available from the manufacturer.

Suddenly applied load

The typical behavior of the generator voltage for a suddenly applied
load, as starting a large motor, is shown in Fig. 5.9. When the load is
applied, the voltage drops initially by ΔV_t; the voltage drops further to
$\Delta V_{t,\,max}$ until the voltage regulator takes control. The voltage finally
recovers with an overshoot and returns to nominal value in a specific
time T_s. The extent of the voltage dip and the recovery time are set by
the parameters and adjustment of the overall excitation and regulation system. The initial dip ΔV_t can be calculated by representing the
generator in Fig. 5.9a by its subtransient reactance x''_d and the motor
by its locked-rotor impedance x_{lr}. The maximum dip $\Delta V_{t,\,max}$ before
the voltage regulator acts can be calculated in the same way, except
for representing the generator by its transient reactance x'_d.

Rectifier load

Emergency/standby generators are required to supply power to rectifier loads for thyristor (SCR) type of dc motor elevator drives, inputs to

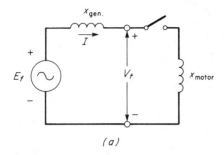

(a)

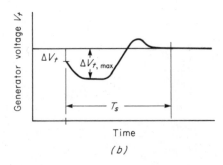

Time

(b)

Figure 5.9   Transient voltage response of generator to motor-
starting load. (a) Equivalent circuit; (b) generator voltage vs.
time.

UPS modules, and battery chargers. The rectifier line current consists of nearly rectangular plus-and-minus 120°-wide pulses as shown in Fig. 5.10b. The current produces distortion of the generator terminal voltage as shown in Fig. 10a. Data processing equipment operating from the same generator can be disrupted by such voltage distortion.

The rms harmonic voltage at the generator terminals for a 500-kVA generator supplying six-pulse thyristor-type dc motor elevator drives is shown in Fig. 5.11. The generator is represented by its subtransient reactance x''_d. Three cases are shown for $x''_d = 0.25$ pu, $x''_d = 0.25$ pu with an isolation transformer for the drive, $x''_d = 0.125$ pu with a transformer. For a limit of 5 percent rms voltage harmonics, the 500-kVA generator can be loaded only to about 50 hp of drives for the first case, 75 kVA for the second, and 125 hp for the third. If the generator is supplying only thyristor drives, lighting and heating loads, and induction motors, the allowable rms harmonic voltages can be increased above 5 percent.

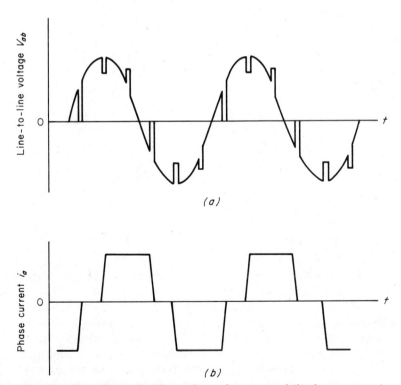

Figure 5.10   Waveforms of (a) line-to-line voltage v_{ab} and (b) phase current i_a of generator under rectifier load.

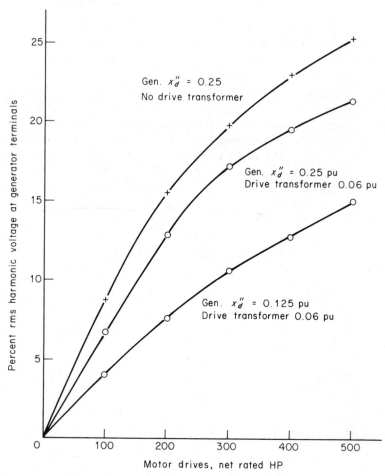

Figure 5.11 Calculated percent rms harmonics at terminals of 500-kVA generator supplying six-pulse SCR rectifier dc drives operating at rated load and speed.

5.6 Generator Protection

Generators require protection against internal and external electrical faults and other conditions that can damage the generator and/or engine under two types of operation. In the first, the generator is operating continuously or on standby independently of the utility line. In the second, the generator is paralleled with the utility line and is operating as a cogenerator or as a peak shaver. The protective system utilizes relays which sense the generator currents and voltages, a neutral grounding resistor or reactor to limit the fault current, and a circuit breaker for connecting the generator to its load. The design of all three elements must be coordinated within the generator and with the external electrical system to ensure the safety of the generator.

Standby operation

The configuration of a 480Y/277-V standby generator is shown in Fig. 5.12 [5]. The generator supplies its load when the transfer switch is placed in the alternate source position. A low-impedance path to the generator neutral is needed here to supply an unbalanced three- or one-phase load. The NEC® requires that this neutral conductor be grounded, since it is used as a service conductor.

To meet these double requirements, the neutral of the generator must be either solidly grounded or grounded through a low reactance. Per NEMA standards, the generator windings are braced to withstand a solid three-phase fault but not a line-to-ground fault. To limit the ground fault current to the three-phase fault level, the neutral is grounded through a low (ohmic) reactance. Normal four-wire operation is possible because the low reactance adds little to the voltage drop for one-phase loads. An air core reactor is used to avoid saturation by the fault current. In the event of a ground fault, it prevents overvoltages from occurring in the unfaulted phases.

As shown in Fig. 5.12, ground fault protection is provided by the overcurrent relay 51N in the neutral and the differential relays 87 in the winding circuits of the generator. For all ground faults in the generator windings, the differential relays trip the generator breaker without delay; the relay 51N acts as a backup to the differential re-

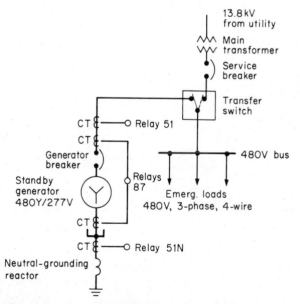

Figure 5.12 Neutral grounding and ground fault protection of a standby generator [5].

lays. For external ground faults, the overcurrent relay 51N trips the breaker. The pickup of this relay is set just above the expected maximum neutral current produced by one-phase or unbalanced loads. Relay 51 trips the generator breaker for external faults not involving ground and for overloads.

Cogeneration

The configuration of a generator that can operate in parallel with the utility line is shown in Fig. 5.13 [5]. The generator can operate continuously to supply building load, sell energy to the utility, and serve as an emergency generator. In most cases the generator is not rated to carry the full building load if the utility supply is lost. Therefore, when the service breaker at the building trips, the generator breaker is transfer-tripped. An exception to this would be when the generator is required to supply the building emergency loads during a utility power failure.

To protect their own systems, most utilities have now formulated detailed protective relay requirements for cogenerators. Utility primary feeder breakers in their distribution substations usually operate on a reclosing cycle. For this reason, the utilities require that the cogenerator also sense a fault in the utility primary feeders and promptly trip its own breaker so that the utility primary feeder break-

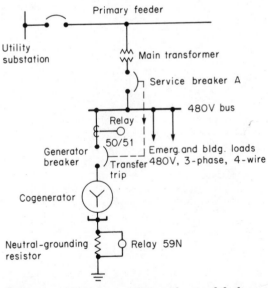

Figure 5.13 Neutral grounding and ground fault protection of a cogenerator [5].

ers can successfully open to clear the fault and then reclose. To do that, the cogenerator must have a phase fault and ground fault protective system matched to the utility protective system.

There are two options for grounding the cogenerator neutral:

Case 1. If the generator is not required to supply any four-wire loads, the neutral can be grounded through a high resistance to protect the generator winding from ground faults.

Case 2. If the generator is required to supply substantial one-phase or unbalanced emergency load, the neutral can be either solidly grounded or grounded through a low reactance as in the case of the standby generator.

Case 1 is shown in Fig. 5.13. Relays 50/51 are the instantaneous and inverse time overcurrent relays for protection against phase faults. Relay 59N is the ground fault overvoltage relay connected across the neutral grounding resistor. When a ground fault occurs within the 480-V system, the service breaker A trips and then transfer trips the generator breaker. When a ground fault occurs in the utility feeder, the utility ground fault relays trip the substation breaker. The fault current supplied by the cogenerator is limited by the neutral grounding resistor. The relay 59N senses the ground fault and trips the generator breaker. The fault is then cleared and the substation breaker can reclose.

In case 2, when the generator neutral is solidly grounded or grounded through a low reactance, the ground fault protection is as shown in Fig. 5.12.

Static relays

Conventional induction disk, induction cup, and balanced beam electromagnetic relays have served the utility industry for several decades, but there are many limitations in their design, application, and maintenance [6]. Today they have been almost completely replaced by reliable solid-state relays. A static relay, which contains no moving parts, is built in an enclosure that is interchangeable with that of an electromagnetic relay. In addition, a static relay has the following advantages over an electromagnetic relay.

- Faster dynamic performance
- Very low burden on current and voltage transformers
- Wider and more flexible frequency characteristics
- Can construct any time/current curve
- Free from overshoot

- Better operation indicators
- Reduced panel space
- Better seismic performance

Two grades of static protective relays are made: utility and industrial. There is no real standard to differentiate between the two grades. The utility-grade relay is a heavier and costlier piece of equipment; the trip setting can be made more accurately, and the relay is usually in a draw-out case. The industrial-grade relay is a packaged solid-state circuit with dials to adjust the settings and LED operation indicators. It is usually not available in a draw-out case, since there are no spring tensions to adjust, no contacts to replace, and no other user-serviceable parts. The package may have edge connectors of the type normally used with PC boards with ratings up to 5 A. The PC board can be arranged with auxiliary relays so that, when it is withdrawn, the CT secondary terminals are automatically shorted. Each of these features reduces the cost of static relays.

5.7 Engines

Engines for driving generators include internal-combustion engines, external-combustion engines, gas turbines, rotary engines, and free-piston engines using either gaseous fuels or liquid fuels or combinations of the two. Capacities of commonly used engine-driven sets range from 5 kW (6.25 kVA) to 2000 kW (2500 kVA). Gasoline engines are used in the lower ranges, and diesel engines are used for sets rated 100 kW or more. Natural-gas- and LP-gas-driven engines are generally available in sizes similar to those of diesels. The requirements for engines are given in NFPA 37 and NFPA 110.*

Fuel systems

A typical fuel system installation includes a day tank and a bulk storage tank. A day tank, mounted on or near the engine, provides a supply of fuel at a relatively constant level, regardless of the level of the bulk storage tank. The day tank does not necessarily hold a day's supply of fuel. Usually, a day tank is specified to hold at least the amount of fuel that would be burned in one hour at rated load. When the bulk storage fuel level is far enough above the day tank to give adequate

---

Reprinted with permission from NFPA 110-1985, Emergency and Standby Power Systems, Copyright© 1985, National Fire Protection Association, Quincy, MA 02269. This reprinted material is not the complete and official position of the NFPA on the referenced subject which is represented only by the standard in its entirety.

gravity flow, a float valve or float switch and fuel shutoff solenoid can be used to refill the day tank. Otherwise, a fuel transfer pump is used.

The capacity of the bulk storage tank is essentially determined by the expected length of a disaster which might interrupt fuel delivery. A commonly used capacity is 14 days of full-load operation. Code exceptions for emergency systems sometimes permit the use of off-site fuel supplies for natural gas and LP gas engines when there is a low probability of both the normal electric utility supply and the off-site gas line failing simultaneously.

Governors

The governor regulates the amount of fuel delivered to the engine at various loads to keep the speed or frequency relatively constant. The mechanical governor uses rotating flyballs to sense speed and operate against a reference spring to actuate the throttle. Higher-performance governors use hydraulic pressure or electric actuators to operate the engine throttle or fuel rack.

The characteristics of a mechanical droop governor are shown in Fig. 5.14. For a setting of the governor, the frequency droops with load. The governor on setting A will provide 60-Hz generator voltage at no load and 59.5 Hz at 50 percent load. When it is reset to C, it will provide 60-Hz generator voltage at 100 percent load.

Electric and electrohydraulic governors are available; they sense speed from the rate of pulses detected by a magnetic pickup facing a toothed gear wheel on the engine shaft. Electric governors are available with the following typical features:

1. Electric load sensing for isochronous load sharing among paralleled sets, i.e., frequency regulation within ± 0.25 percent from no load to full load

2. Electric load sensing for load anticipation to actuate the throttle

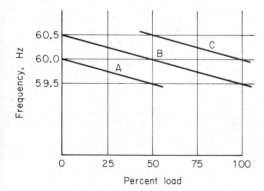

Figure 5.14 Governor characteristic.

before the engine speed and generator frequency drop to reduce response time

3. Ramp acceleration to reduce frequency overshoot on start-up

4. Electric speed adjustment for automatic synchronizing of generators

5. Monitors for maximum engine power and to prevent reverse power flow into the generator

Starting

Engines are started by using battery-powered electric motor starters or compressed-air motors, depending upon the type of engine and the availability of compressed air. Electric starting is the most common method. The starting batteries should have sufficient capacity for 60 s of continuous cranking. The engine manufacturer defines the battery capacity in cold-cranking amperes. Lead-acid or nickel-cadmium batteries are used. The batteries are normally maintained and recharged by a battery charger supplied from the emergency system.

When compressed air is available, the engine can be started by an air motor or, in larger engines, by direct air injection into the cylinders. An air tank with sufficient capacity to provide five 10-s cranking attempts is usually required. The air compressor for refilling the tanks during a normal utility outage can be driven by an electric motor from the emergency system or by an auxiliary gasoline engine.

5.8   Controls

Both the engine and the generator require panels for displaying operating conditions, mounting the controls, providing protection, and displaying alarms.

Engine panel

The engine control panel provides the following typical functions:

1. Cranking control equipment, including cranking cycle, battery charger state, and battery condition

2. Control switches, marked RUN–OFF–AUTOMATIC

3. Shutdown and lock-out controls, to operate from the protection system

4. Annunciator, to respond to protection system

5. Emergency shutdown

6. Governor controls, including raise–lower speed

7. Instruments, including oil pressure and water temperature

Engine protection

The safety indications and protection for the engine are shown in Fig. 5.15.

Table 3-5.5.1(d)   Safety Indications and Shutdowns

Indicator Function (at Battery Voltage)	Level 1 C.V.	Level 1 S	Level 1 R.A.	Level 2 C.V.	Level 2 S	Level 2 R.A.
(a) Overcrank	X	X	X	X	X	O
(b) Low Water Temp. < 70° F (21° C)	X		X	X		O
(c) High Engine Temperature Prealarm	X		X	O		
(d) High Engine Temperature	X	X	X	X	X	O
(e) Low Lube Oil Pressure Prealarm	X		X	O		
(f) Low Lube Oil Pressure	X	X	X	X	X	O
(g) Overspeed	X	X	X	X	X	O
(h) Low Fuel Main Tank	X		X	O		O
(i) EPS Supplying Load	X		O			
(j) Control Switch Not in Auto. Position	X		X	O		
(k) Battery Charger Malfunctioning	X		O			
(l) Low Voltage in Battery	X		O			
(m) Lamp Test	X		X			
(n) Contacts for Local and Remote Common Alarm	X		X	X		X
(o) Audible Alarm Silencing Switch			X	O		
(p) Low Starting Air Pressure	X		O			
(q) Low Starting Hydraulic Pressure	X		O			
(r) Air Shutdown Damper when Used	X	X	X	X	X	O
(s) Remote Emergency Stop	X			X		

Key:
C.V.   Control panel-mounted visual   S Shutdown of EPS
       indication   X Required
R.A.   Remote Audible   O Optional

Notes:
1. Item (n) shall be provided, but a separate remote audible signal shall not be required when the regular work site of 3-5.6.1 is manned 24 hours per day.
2. Item (b) is not applicable for combustion turbines.
3. Items (p) or (q) apply only when applicable as a starting method.
4. Item (i) EPS AC Ammeter is acceptable for this function.
5. All required C.V. functions visually annunciated by a remote, common visual indicator. All required functions indicated in Column R.A. also annunciated by a remote, common audible alarm [see 3-5.5.1.(d)].

Figure 5.15   Safety indications and shutdowns for engine protection. Table 3-5.5.1(d) of NFPA 110 [1]. Reprinted with permission from NFPA 110-1985, Emergency and Standby Power Systems, Copyright©1985, National Fire Protection Association, Quincy, MA 02269. This reprinted material is not the complete and official position of the NFPA on the referenced subject which is represented only by the standard in its entirety.

Generator panel

The generator control panel provides the following typical functions:

1. An ac voltmeter with a phase selector switch
2. An ac ammeter for each phase, or phase selector switch
3. A frequency meter
4. Voltage adjustment for the voltage regulator
5. Protective relays

Generator protection

The generator is protected by the following typical devices:

1. Phase overcurrent relays
2. A differential relay for each stator winding
3. A reverse power relay
4. A ground fault relay, in the neutral circuit

5.9 Installation

NFPA 37 [3] governs the fire safety for the installation and operation of stationary combustion engines and gas turbines not exceeding 7500 hp per unit. The standard applies also to portable engines which remain connected for use in the same location for a period of one week or more and which are used instead of or to supplement stationary engines. Engines used in essential electrical systems in health care facilities shall comply with this standard and any special provisions in NFPA 99 [7].

Location

Engines, with or without their weatherproof housings attached to the engine subbase, may be installed outdoors, inside structures, or on the roofs of structures. Engines which are installed outdoors or on the roofs of structures shall be located at least 5 ft from openings in walls and at least 3 ft from structures having combustible adjacent walls. Engines rated more than 50 hp shall be located outdoors, on roofs, in special detached structures, or in rooms within or attached to other structures.

Detached structures shall be of noncombustible or fire-resistant construction. Rooms located within structures shall have interior walls, floors, and ceilings of at least 1 h fire resistance rating. Rooms attached to structures shall meet the requirements of detached structures. Stationary engines shall be supported on firm foundations or suitable steel framework properly secured. Provision shall be made to supply sufficient air for combustion, proper cooling, and adequate ventilation.

Fuel supply

Engines can be supplied by fuel gas, LP gas in the liquid phase, gasoline, or diesel oil. Gas piping shall be installed in accordance with NFPA 54 [8] for fuel gas 60 psi and less, NFPA 58 [9] for fuel gas greater than 60 psi, and for LP gas in the liquid phase. Every gas engine shall have a carburetion valve, zero governor-type regulating valve, fuel control valve, or an auxiliary valve which will automatically shut off the flow of gas if the engine stops for any reason. For gas-fueled engines, only integral tanks shall be permitted inside structures or on roofs. An integral tank shall not exceed 25 gal capacity. It shall be mounted on the engine assembly and be protected against vibration, physical damage, engine heat, and the heat of exhaust piping. Tanks other than integral tanks shall be located underground or above ground outside structures.

For diesel-oil-fueled engines, only one securely mounted integral tank shall be installed on each engine. Unenclosed day tanks and supply tanks installed in the lowest story, cellar, or basement supplying engines which drive generators, alternators, fire pumps, or other equipment used for emergency purposes shall not exceed 660 gal. Larger tanks shall be enclosed within walls, floor, and top having a fire resistance of not less than 3 h.

Exhaust

Engine exhaust discharge systems shall be designed on the basis of flue gas temperatures: low-heat appliances for temperatures not exceeding 1000°F; high-heat appliances, above 1000°F. Exhaust pipes shall be of wrought iron or steel. Provision shall be made in exhaust systems to prevent damage resulting from the ignition of unburned fuel. When necessary, a flexible connector shall be provided in the exhaust pipe from the engine to minimize the possibility of a break in the engine exhaust system because of engine vibration or heat expansion. The exhaust pipe shall terminate outside the structure at a point where the hot gases or sparks will be discharged harmlessly and not be directed against combustible material or structures or into atmosphere containing flammable gases or vapors or combustible ducts.

Example 5.1   Engine-generator. A standby engine-generator set is to supply a maximum load of 1000 kVA with 10 percent margin. Included in the load is an induction motor rated 600 hp, 0.85 PF, 0.90 full-load efficiency. The engine-generator set cannot start an induction motor rated more than 50 percent of the set rating without causing excessive voltage dip. Select a maximum size engine-generator set from units rated 1000, 1250, and 1500 kVA.

solution   The set size based on maximum load is

$$\text{Rating} = 1000 \text{ kVA} \times 1.1 = 1100 \text{ kVA}$$

The set size based on motor starting is

$$\text{Motor rating} = \frac{600 \text{ hp} \times 0.746 \text{ kW/hp}}{0.85 \text{ PF} \times 0.90 \text{ eff.}} = 585 \text{ kVA}$$

$$\text{Rating} = 585 \text{ kVA}/0.5 = 1170 \text{ kVA}$$

The minimum rating to satisfy the motor-starting requirement is 1250 kVA.

Example 5.2   *Motor starting*. Find the maximum horsepower rating induction motor that can be line-started from the following generator without exceeding 15 percent voltage dip.

Generator   500 kVA, impedance $x_d'' = 12$ percent
Motor $6 \times$ starting current, 0.85 PF, 0.92 eff.

solution   The voltage dip in terms of motor input kVA is

$$V \text{ dip percent} = (\text{motor kVA}/500 \text{ kVA}) \times 6 \times 12$$
$$= 0.144 \times \text{motor kVA}$$

The motor kVA for 15 percent dip is found by

$$\text{Motor kVA} = 15/0.144 = 104 \text{ kVA}$$

The motor hp rating is found by

$$\text{Motor hp} = \frac{104 \text{ kVA} \times 0.85 \text{ PF} \times 0.92 \text{ eff.}}{0.746 \text{ kW/hp}}$$

$$= 109 \text{ hp, say, } 100 \text{ hp}$$

5.10   Summary

Sets using rotating equipment, i.e., emergency power supplies, include generators driven by engines, gas turbines, and steam turbines. The sets are fueled with liquid petroleum, liquified petroleum gas, and natural or synthetic gas. Diesel-engine-driven generator sets ranging in size from 100 to 2000 kW predominate in standby/emergency systems. The sets are used singly or in parallel groups up to four units. Installation requirements are covered in NFPA codes.

REFERENCES

1. ANSI/NFPA 110-1985, "Standard for Emergency and Standby Power Systems."
2. ANSI/NFPA 70-1987, "National Electrical Code®."
3. ANSI/NFPA 37-1984, "Standard for the Installation and Use of Stationary Combustion Engines and Gas Turbines."
4. A. E. Fitzgerald, C. Kingsley, Jr., and S. Umans, *Electric Machinery,* 4th ed., McGraw-Hill, New York, 1983.
5. A. Kusko and S. M. Peeran, "Neutral-Grounding and Protection for Energy-Saving Generator Sets," *Engineer's Digest,* April 1987, p. 64.
6. A. Kusko and S. M. Peeran, "Minimize Cost of Engine-Generator Protection with Static Relays," *Power,* May 1987, pp. 35–39.
7. ANSI/NFPA 99-1984, "Standard for Health Care Facilities."
8. ANSI/NFPA 54-1984, "National Fuel Gas Code."
9. ANSI/NFPA 58-1986, "Standard for the Storage and Handling of Liquefied Petroleum Gases."

Chapter

6

Static UPS

The static uninterruptible power system (UPS) provides the following functions for computers and other critical loads:

1. Buffers the load from the line. Eliminates power line noise and voltage transients from the load.
2. Provides voltage regulation for any level of line voltage, including brownouts.
3. Protects load against line-frequency variations.
4. Converts power to higher frequency, e.g., 60-to-415-Hz conversion.
5. Provides uninterruptible power during failures of the normal utility line.

The UPS is assembled from modules, each typically rated up to 600 kW (750 kVA) for 60-Hz output and 169 kW (187.5 kVA) for 415-Hz output. Batteries are selected to obtain up to 20 min of operation; engine-generator sets are usually provided to extend the running time beyond the battery capacity.

The categories of load in systems supplied by UPS are usually defined as follows:

1. Critical, may not be interrupted for more than one-half cycle (of 60-Hz power) supplied by UPS.
2. Essential, may not be interrupted for more than 10 s; supplied by UPS or engine-generator.
3. Nonessential, may be interrupted for the duration of a normal utility line failure.

73

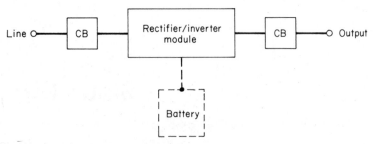

Figure 6.1   Block diagram of single-module UPS.

6.1   Standards

Particular aspects of UPSs are covered in the following:

1. ANSI/NFPA 70-1987, "National Electrical Code®,"* Section 700-12(t) [1]

2. ANSI/IEEE Std. 446-1987, "IEEE Recommended Practice for Emergency and Standby Power Systems for Industrial and Commercial Applications," Chap. 5, Stored Energy Systems [2]

The power-semiconductor components, rectifier, and inverter subassemblies of UPS are described by Dewan and Straughen [3].

6.2   Application

Static UPSs are used in at least four configurations, as shown in Figs. 6.1 to 6.4.

Figure 6.1 shows a single-module UPS. The input is 60 Hz; the output is either 60 or 415 Hz. Single-module UPSs range from several hundred watts to several hundred kilowatts. Modules with self-contained batteries for installation adjacent to a computer typically range up to 50 kW. Larger modules use battery cabinets or separate battery racks.

Figure 6.2 shows a single-module UPS with bypass. The input and output must be the same frequency, i.e., 50 or 60 Hz. Operation on the bypass line takes place by manual selection or from automatic transfer when the module fails. Transfer to bypass may also take place when the module sees an overload or a fault in its output circuit. Various switch arrangements are used to shunt the solid-state switch and to isolate the equipment for maintenance.

---

*National Electrical Code® and NEC® are registered trademarks of the National Fire Protection Association, Inc., Quincy, Mass.

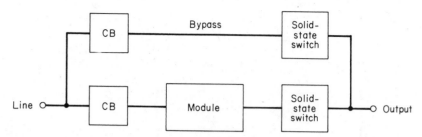

Figure 6.2   Block diagram of a single-module UPS with bypass.

Figure 6.3 shows a three-parallel-module configuration. The additional modules are used to obtain more power and to increase the reliability. When an extra module is redundant, any one module can be tripped from the system; the remaining modules carry the load. A bypass switch can be used with the system to provide additional reliability. All of the modules run from a common battery.

Figure 6.4 shows a two-module configuration with a standby generator. The battery capacity for a UPS is typically sized to carry full load for up to 20 min. For longer-term utility outages, an engine-generator set is required to pick up the input bus, supply the inverters, and recharge the batteries. The engine-generator also supplies essential load.

6.3   UPS Module

Each UPS module, regardless of power rating, includes the three sections shown in Fig. 6.5. The rectifier section converts the input ac power to dc power to charge the battery and supply the inverter section. The dc voltage is regulated in the face of variation of the input voltage and load to maintain the battery at the desired float charge

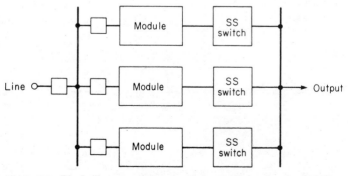

Figure 6.3   Block diagram of three-module parallel-redundant UPS.

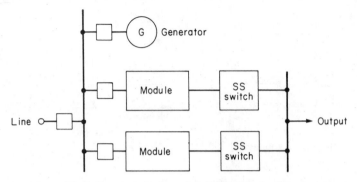

Figure 6.4 Block diagram of two-module UPS with standby generator.

condition or at a specific voltage per cell. Some modules are built with a separate battery charger, so that the main rectifier does not have to be regulated. In that case, the battery is connected by a high-speed static switch to the dc bus when the input power fails. The rectifier section will be treated in more detail in Sec. 6.4.

The dc section (Fig. 6.5), also termed the dc link, includes the battery, battery switches, dc filter inductor, filter capacitors, charging circuit, and dc circuit breaker. As previously explained, the batteries are connected either directly to the bus or through a high-speed switch. Up to about 50 kW, the batteries may be mounted in the same cabinet as or in a cabinet adjacent to the module. Above that module rating, the batteries are mounted in racks in a battery room. The battery may be dedicated to one module, may be common to all modules in a parallel system, or may be common to all modules for a 60- and 415-Hz UPS at the same site.

The inverter section (Fig. 6.5) converts the dc power to ac output power at the desired frequency, voltage, and waveform. The inverter

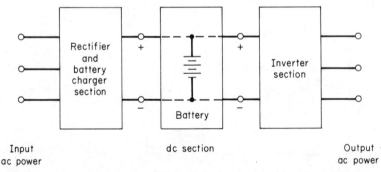

Figure 6.5 Block diagram of a UPS module showing rectifier, dc, and inverter sections.

synthesizes the output voltage by means of a group of solid-state switches (transistors or thyristors) whose operation will be described more fully in Sec. 6.5. The section also includes the filters for smoothing the output waveform and switches for connecting the section to the output bus.

In addition to the power-handling components described above, the module includes control circuits, logic circuits, and indicating instruments for operation and monitoring. It also requires cooling fans, air filters, and other hardware. All of the auxiliaries must be able to operate when it is operating from either the normal ac input or the battery. Modules intended for operation in parallel-redundant configurations must also include circuits for synchronizing, load division, and reactive-power division among themselves.

6.4   Theory of Operation: Rectifier

In the rectifier section, ac power is converted to dc power by means of either diodes or thyristors, as shown in Fig. 6.6. The diode conducts current when voltage is applied from anode to cathode, $+v_d$, and blocks current when the voltage is $-v_d$. In the conducting direction, the voltage drop in the diode is 1.5 to 2.5 V at its rated current. The voltage rating of the diode is a measure of its ability to block reverse current. The thyristor shown in Fig. 6.6b has an extra terminal termed a "gate." Like the diode, the thyristor blocks current for $-v_t$. However, the thyristor will not conduct current in the forward direction, $+v_t$, until a voltage signal is applied to the gate. By shifting the voltage signal in phase with respect to the ac voltage applied to the thyristor, the conduction period can be controlled and the dc output voltage of the rectifier can be controlled as well.

The commonly used three-phase six-pulse bridge rectifier circuit is shown in Fig. 6.7. The bridge is supplied with three-phase power through a set of rectifier transformers which serve to isolate electrically the dc side of the rectifier from the input ac line and to set the dc

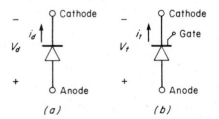

(a)   (b)

Figure 6.6   Semiconductors for rectifiers. (a) Diode; (b) thyristor (SCR).

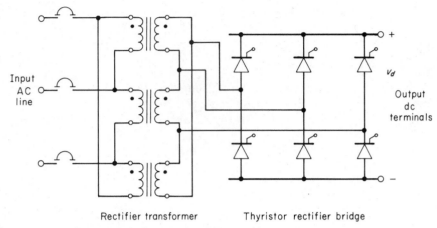

Rectifier transformer   Thyristor rectifier bridge

Figure 6.7   Schematic diagram of six-pulse thyristor rectifier.

voltage level of the output terminals. The six semiconductor devices can be thyristors, as shown, to permit control of the dc output voltage, or diodes, which deliver only the maximum output voltage. The term "six-pulse" means that the output voltage ripple has six pulses for each cycle of the ac input voltage.

Sample waveforms of the dc output voltage of the rectifier circuit of Fig. 6.7 are shown in Fig. 6.8. One cycle spanning 360° of the ac input voltage is shown. During each 60° interval, the highest instantaneous line-to-line voltage is connected through one device in the upper group and one device in the lower group to the dc output terminals. Hence, the ripple shown in Fig. 6.8a is the top 60° of the line-to-line voltage. That waveform is produced when the devices are diodes or when they are thyristors with a firing angle of $\alpha = 0°$. When the firing angle is increased (delayed), the thyristors also conduct in 60° intervals but with notches as shown, for example, in Fig. 6.8b for $\alpha = 30°$. The average voltage V_d is lower than that for the diode rectifier in Fig. 6.8a.

The average output voltage V_d as a function of the firing angle α is shown in Fig. 6.9. At $\alpha = 0°$, V_d corresponds to the diode operation; at $\alpha = 90°$, $V_d = 0$. However, the instantaneous output voltage has considerable, but equal, plus-and-minus ripple. Single-phase rectifiers in low-power modules use two or four diodes or thyristors to produce two-pulse ripple, but the control characteristics are similar.

6.5   Theory of Operation: Inverter

A three-phase inverter employs a minimum of six semiconductor switches to synthesize the ac output voltages from the dc input voltage. The switches are transistors, thyristors, or gate turnoff devices

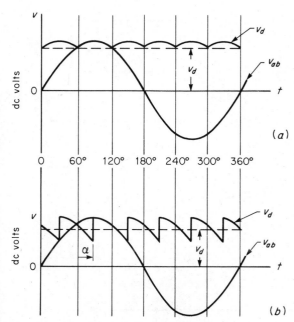

Figure 6.8 Waveforms of six-pulse thyristor rectifier of
Fig. 6.7: dc output voltage v_d and average voltage V_d for
(a) firing angle $\alpha = 0°$ and (b) firing angle $\alpha = 30°$.

(GTOs). The switches must be closed and opened in a specific sequence and time duration. The common modes of operation are termed "six-step" and "pulse-width modulation (PWM)."

Six-step inverter

The sequence in which the thyristors are operated to obtain three-phase output waveforms is shown in Fig. 6.10 [4]. The thyristors are represented by switches, and the load is represented by a Y-connected

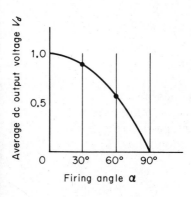

Figure 6.9 Six-pulse thyristor rectifier of Fig. 6.7. Average dc output voltage V_d vs. firing angle α.

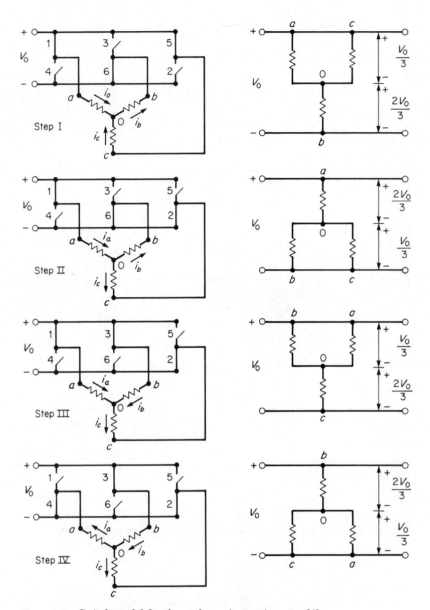

Figure 6.10 Switch model for three-phase six-step inverter [4].

set of impedances. Each step corresponds to 60° on the output waveform. During step I, thyristors 1, 6, and 5 are conducting. The a and c ends of the load are connected to the positive dc supply bus, and terminal b is connected to the negative supply bus. The line voltages

V_{ab} and V_{bc} are each V_0 in magnitude. The line-to-neutral voltages V_{a0} and V_{c0} are each $V_0/3$, and V_{b0} is $-2V_0/3$, as shown in Fig. 6.10. At each 60° interval, one of the thyristors switches in the sequence 1-2-3-4-5-6. The waveforms for the line-to-neutral voltages V_{a0} and V_{b0} and the line voltage V_{ab} are shown in Fig. 6.11. The line-to-neutral voltages are six-step-per-cycle approximations to sine waves, and the line-to-line voltages are 120°-wide positive and negative pulses. The voltage waveforms are characterized by a fundamental component at

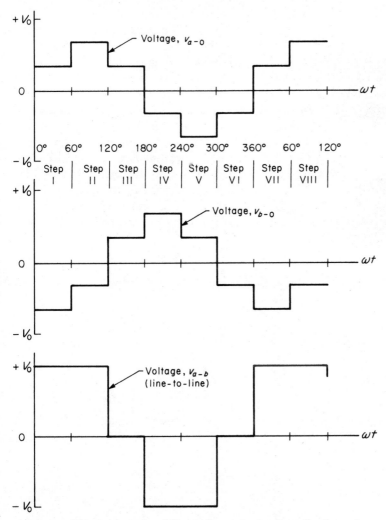

Figure 6.11   Line-to-neutral and line-to-line output voltage for three-phase six-step inverter [4].

the desired frequency, e.g., 60 or 415 Hz, and harmonics of the order 5th, 7th, 11th, 13th, etc. The waveforms are "smoothed" by using filters or other cancellation techniques to remove the harmonics.

Commutation

The thyristor, as a semiconductor switch, is turned on by a gate pulse; it requires an externally applied current through the cathode-to-anode terminals to turn it off. The process is called commutation. The GTO can be turned off by a negative gate pulse. The transistor is turned on and off by applying current to the base. The commutation process for the thyristor inverter will be described.

An electric circuit model for the upper and lower legs of an inverter, e.g., 1 and 4 of Fig. 6.10, is shown in Fig. 6.12. In addition to the main thyristor switches T1 and T2, the circuit requires commutating thyristors T3 and T4, reactive diodes D1 and D2, and commutating components C and L. The function of the additional components is to transfer the conduction from thyristor T1 to T2 and back at the correct times. Prior to the switching time, the current I_1 is carried by T1 to the a-phase terminal of the load. At the required time, thyristor T3 is fired. The capacitor C discharges with current I_2 through diode D1; the forward drop reverse-biases thyristor T1 until it turns off. The load current I_3 transfers to diode D2 until the current reaches zero.

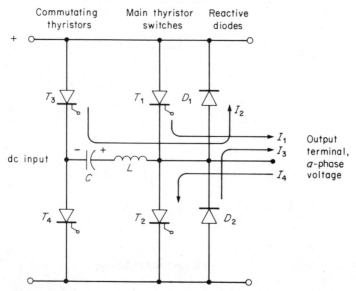

Figure 6.12 Circuit for the upper and lower legs of an inverter showing the commutation of the current from T_1 to T_2.

Thyristor T2 is fired to carry the load current I_4 in the negative direction. The switching cycle is completed.

During the time that diode D2 is carrying the load current I_3, energy is transferred from the load back to the dc source for the inverter. That energy is stored in filter capacitors and is returned to the load when thyristor T2 starts to carry load current I_4.

Pulse-width modulation

An alternative way to operate the legs of an inverter is to operate each solid-state switch multiple times each half cycle in the pulse-width modulation (PCM) mode. An example of such switching six times per half cycle is shown in Fig. 6.13 [3]. The switching points are generated by sampling a sine wave with a sawtooth wave so that the pulse widths and spacings approximate the sine wave. The solid-state switches must be capable of switching on and off at a frequency higher than the UPS output frequency, e.g., 6 times 60 or 415 Hz. Thyristor switches are usually not used because of their turn-off time and both switching and commutation circuit losses. Transistor switches are used for PWM inverters up to several hundred kilowatts, and GTOs are used for higher power.

PWM operation results in several size and cost reduction features. First, the dc bus can operate at constant voltage from diode rectifiers; voltage control is effected by controlling the width of each pulse. Second, the output filter is smaller than for six-step operation; the lowest frequency to be filtered is $2N$, where N is the number of pulses per half cycle.

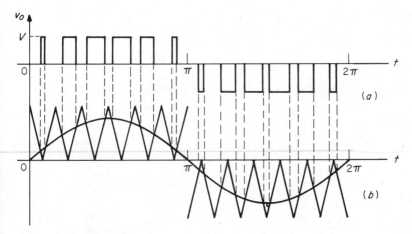

Figure 6.13 Voltage waveform under pulse width modulation; $N = 6$.

Output voltage

In an actual UPS module, the inverter voltage as shown in Fig. 6.11 must be regulated and filtered before it can be applied to the load. The amplitude of the ac voltage is proportional to the input dc bus voltage, which is usually fixed to the battery voltage. Several methods are used to regulate voltage; the following are examples.

1. *Phase-shift control.*   Two inverters are used; their output voltages are connected in series. By adjusting the relative phase angle of the two voltages, the sum can be adjusted to control the net output voltage.

2. *Pulse-width control.*   The width of each step in Fig. 6.11 is adjusted from 60° down. The net voltage is thus controlled.

3. *Pulse-width modulation.*   The inverter is turned on and off many times during each step in Fig. 6.13. By adjusting the ratio of on time to off time, the net voltage is controlled.

The inverter voltage is described as the sum of the desired fundamental-frequency component (60 Hz) and undesired harmonics. The objective of generating a sine wave voltage at the output terminals of the module is achieved in two steps. First, the inverter is designed to produce a voltage waveform that is close to sinusoidal, e.g., 12 steps instead of 6 steps. Second, the residual harmonic voltages are suppressed by means of filters. Figure 6.14 shows an example of the one-phase circuit of a filter. The voltage v_1 from the inverter contains the 60-Hz fundamental plus 5th-harmonic (300-Hz) and 7th-harmonic (420-Hz) components. In the filter, L_1 and C_1 are tuned to be series-resonant at 300 Hz; L_2 and C_2 are tuned to 420 Hz. Each filter section presents a low impedance to its corresponding harmonic so that its harmonic voltage is dropped in the series inductor L_s and does not reach the output voltage v_2.

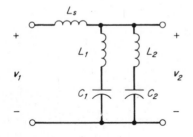

Figure 6.14   One phase of a typical power filter section.

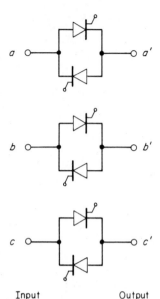

Figure 6.15 Circuit of a solid-
state three-phase ac switch.

Input   Output

Static switch

A three-phase static switch consists of six thyristors connected back-
to-back in pairs as shown in Fig. 6.15. The static switches are used for
at least three functions: (1) to connect (and disconnect) a UPS module
from a common output bus, (2) to connect a standby module to the
load, (3) to connect a bypass line to the load. A static switch can open
or close in less than 4 ms, compared to at least 30 ms for an electromechanical switch.

To close the switch, gate signals are applied to all six thyristors; the
individual devices turn on when subjected to positive anode-to-cathode
voltage and off when the voltage is negative to carry the three-phase
current in the manner of an electromechanical switch. To open the
switch, either the gate signals can be removed and the conducting devices allowed to naturally commutate off in a matter of milliseconds,
or the conducting devices can be force-commutated to terminate conduction more rapidly.

6.6 Nonlinear Load

Switching power supplies are used in computers, peripherals, and
communications equipment to provide low dc voltages for logic circuits and microprocessors and to meet other requirements. Each of
these power supplies consists of a rectifier, filter capacitor, high-
frequency switch (e.g., 40 kHz), high-frequency transformer, second

rectifier, and filter. They are considerably smaller, cheaper, and more efficient than conventional power supplies with 60-Hz transformers. However, the current from the 60-Hz line to the switching power supply consists of positive and negative pulses because of the peak charging of the filter capacitor. The consequences are line currents like those shown in Fig. 6.16 that must be supplied by the UPS [5].

These currents consist of a fundamental plus odd harmonics, primarily the 3d harmonic. Computer loads are typically supplied by one-phase two-wire and three-phase four-wire circuits, including a neutral. These harmonic currents will produce voltage drops in the internal source impedance of the UPS, as seen in the flattening of the voltage wave shown in Fig. 6.16. The flattening reduces the dc voltages delivered by the power supplies in the facility equipment, and it may initiate low-voltage alarms or shutdowns.

The internal impedance of the UPS may be abnormally high at specific harmonic frequencies, particularly the 3d, because of the output filter. An example of one section of the output filter for a six-step inverter is shown in Fig. 6.17a. The per-unit values of reactance at 60 Hz are typical. The 5th-harmonic section is tuned to be resonant at 300 Hz to remove the 5th-harmonic voltage component from the internal UPS voltage. The filter would also have sections for the 7th, 11th, and 13th harmonics. The output impedance is shown in Fig. 6.17b. The section is parallel-resonant at about the 4th harmonic. The impedance at the 3d harmonic is about 0.5 pu on the inverter base, compared to about 0.15 pu if the power were supplied by a transformer with 0.05-pu impedance at 60 Hz.

A simple *LC* filter for a PWM inverter is shown in Fig. 6.18a designed to attenuate a modulation frequency of 720 Hz (12 × 60 Hz). The output impedance is shown in Fig. 6.18b as a function of frequency. The *LC* combination is parallel-resonant at about the 4th har-

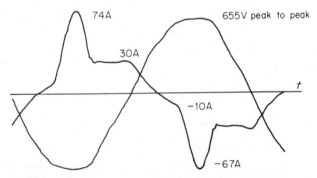

Figure 6.16 Typical input currents of data processing devices. Load: two VAX 11/780 processors [5]. (*©1987 IEEE*)

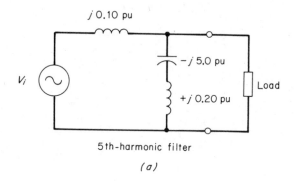

5th-harmonic filter

(a)

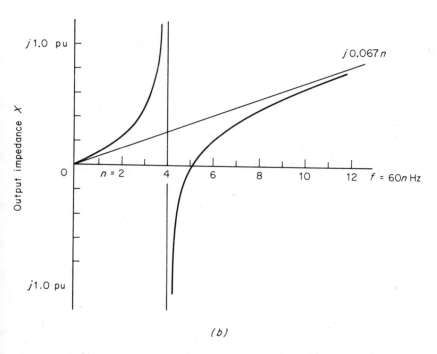

(b)

Figure 6.17 Six-step inverter. (a) Filter section for 5th-harmonic voltage; (b) output impedance as a function of frequency.

monic. The impedance at the third harmonic is about 0.8 pu on the inverter base.

The effect of the distorted line current of the nonlinear loads on the UPS can be reduced in several ways. First, the load should be supplied line-to-line and distributed over the three phases so that the 3d-harmonic current components are suppressed. Second, the percent of

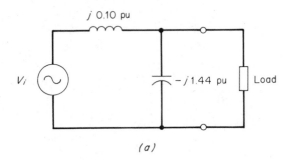

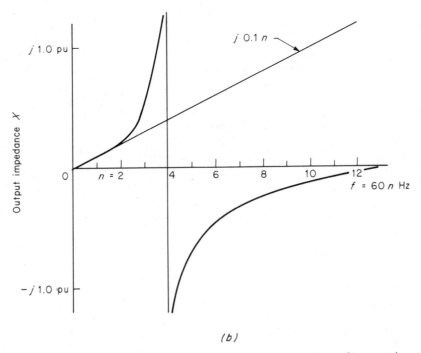

Figure 6.18 PWM inverter. (*a*) Filter to attenuate 720-Hz voltage; (*b*) output impedance as a function of frequency.

nonlinear load supplied by any one UPS module should be minimized. Third, the nonlinear load should be supplied from a rotary UPS; the generator has a relatively low output impedance compared to a static UPS with filter.

6.7   UPS Control Section

A typical floor-standing single-module UPS shall include the following control equipment [6]:

Instrumentation

The system control panel shall be provided with the following metering:

1. Input voltage and current meters with phase selector switch
2. DC battery charge/discharge ammeter
3. DC battery voltage meter
4. AC voltmeter with selector switch to monitor the UPS output and bypass line with the same meter
5. Output and bypass ac ammeter

Controls

The following system level control functions shall be provided on the system control panel:

1. UPS/bypass transfer switch
2. AC output voltage adjustment ± 5 percent
3. Battery circuit breaker trip push button
4. Emergency shutdown push button with protective cover
5. Lamp test/reset push button
6. Audio alarm test/reset push button

Alarms

The system control panel shall be provided with the system level alarm lights listed below. All alarm lights shall be long-life LEDs; incandescent bulbs shall not be used.

1. Overload alarm
2. Overload shutdown alarm
3. Equipment overtemperature alarm
4. Ambient overtemperature alarm
5. Fuse failure alarm

6. Fan failure alarm

7. Battery breaker open alarm

8. Battery discharging alarm

9. Low-battery alarm

10. DC overvoltage alarm

11. Input power-failed alarm

12. Control power-failed alarm

13. Emergency off alarm

14. Output overvoltage/undervoltage alarm

15. Static switch unable alarm

16. Load on bypass

An audible alarm is to come on when any of the above alarms come on.

Remote alarm panel

A wall-mounted remote alarm panel is to be provided and is to include the following:

1. UPS on battery alarm

2. Low-battery alarm

3. Load on bypass alarm

4. Summary alarm for all module malfunctions

5. Audible alarm with reset push button

6. Lamp test/reset push button

Self-diagnostic circuits

Built-in diagnostic circuits for troubleshooting and circuit alignment aids shall provide indication of the following:

1. Rectifier in control mode

2. Module synchronizing with critical load bus

3. Positive dc bus ground fault

4. Negative dc bus ground fault

5. Bypass frequency higher than system output frequency

6. Bypass frequency lower than system output frequency

7. Automatic static transfer switch lockout

8. Command given to close system output circuit breaker

9. Command given to open system output/bypass circuit breaker

10. Command given to open system output/bypass circuit breaker

11. The degree of overload

12. Undervoltage trip for battery circuit breaker

13. Undervoltage trip for module input circuit breaker

6.8 Reliability Improvement

The MTBF of a UPS module is typically 10,000 to 20,000 h. Theoretically, in a system of four modules, a module failure could occur two or four times a year. To reduce the impact of a module failure on the operation of a critical load, two measures are used: a bypass line, as shown in Fig. 6.2, and parallel-redundant modules, as shown in Fig. 6.3. The subject of reliability is treated in more detail in Chap. 14.

The bypass line of Fig. 6.2 can be used only when the module is delivering power at the same frequency as the input line, i.e., 60-Hz output, 60-Hz input. The inverter output voltage is maintained in synchronism with the input voltage. The load power is transferred from the module to the bypass line by means of high-speed solid-state switches in response to either a manual planned signal (push button) or an automatic unplanned signal. The signal is obtained from a group of circuits that monitor the inverter for a failure, so that the transfer to bypass can be made without disturbing the critical load. A second bypass line directly from the utility line to the output bus, called a maintenance bypass, is frequently used. This bypass isolates the module and the static switches for maintenance.

When the system of Fig. 6.3 is assembled with one module more than is required for the load, it is termed a "singly redundant parallel-module system." Any one module can be taken out for planned maintenance or can trip out for a failure, and the remaining modules will carry the load. Interconnecting control lines among the modules are required to maintain synchronism and provide signals for real and reactive power division among the modules. A bypass line is usually incorporated, as well, for the module system to increase reliability and permit maintenance.

6.9 Installation

The installation requirements for UPSs are set by the power rating as follows:

100 W to 10 kW

Typical UPSs in the 100-W to 10-kW power range are shown in Figs. 6.19 and 6.20. The modules include self-contained batteries and can be placed adjacent to the critical load equipment to be supplied.

10 kW to 100 kW

A 30-kVA UPS module is shown in Fig. 6.21. These modules may contain batteries in a separate cabinet mounted adjacent to the inverter module directly on the computer floor. Or, the inverter module and a separate battery rack may be located in an existing electrical room or storage room. The location requires ventilation for the heat generated by the module power losses. Sealed batteries are normally used.

100 kW to 1000 kW, and larger

A 500-kVA module is shown in Fig. 6.22. A typical installation of multiple modules is shown in Fig. 1.2; the battery room is shown in Fig. 6.23 [7]. The multiple modules are located in a UPS room, usually with the input and output switchgear and transformers. The room is

Figure 6.19 UPS unit with supplementary battery. Rating, 1.5 kW; input, 120 V, one-phase; output, 120 V, one-phase; backup time, 15 to 20 min; recharge time, 10 to 12 h. (*Courtesy Emerson Industrial Controls*)

Figure 6.20   UPS units rated 3.5, 10 kVA; input, 120 V, one-phase; output, 120 V, one-phase. Self-contained batteries; backup time, 10 min. (*Courtesy Emerson Industrial Controls*)

Figure 6.21 UPS module rated 30 kVA; input and output, 480Y/120 V. Separate battery cabinet; backup time, 15 min; weight of module, 1300 lb; battery weight, 1700 lb. (*Courtesy Emerson Industrial Controls*)

Figure 6.22   UPS module rated 500 kVA; input and output, 208Y/120 V, 480Y/277 V, or 600 V, three-phase, 60 Hz. Separate battery facility. (*Courtesy Emerson Industrial Controls*)

usually air-conditioned to remove the heat of the losses and to reduce the temperature stress on the electronic components. The batteries are located in a battery room, which requires ventilation in accordance with the NEC®. The interior of one inverter cabinet is shown in Fig. 6.24 [7].

Figure 6.23   Battery facility for UPS installation.

Figure 6.24 Interior of inverter cabinet for module of Fig. 6.22. (*Courtesy Emerson Industrial Controls*)

The installation must be designed to expedite rapid servicing to minimize downtime. Such factors as the following should be considered:

1. Ample space around the modules to work, remove subassemblies such as drawers, and remove heavy components such as transformers

2. Ample space adjacent to the UPS room to store spare parts, manuals, test equipment, and maintenance records

3. Permanent or portable dummy loads to test modules before returning them to service

4. Breaker, switch, and terminal arrangements so that ac and dc high-voltage equipment can be shut down in any compartment that requires servicing

Example 6.1 [8]   Select a single UPS module to supply the following computer room loads. Allow 50 percent of the actual load for additional growth.

1. At 208Y/120 V, three-phase:
    - CPU   2 units, 22 A each
    - CPU   1 unit, 17 A
    - Disk drive   4 units, 9.5 A each

2. At 120 V, one-phase:
   - Printer   4 units, 11 A each
   - Printer   2 units, 3 A each
   - Plotter   6 units, 4 A each
   - Terminals   12 units, 4 A each

solution   The power required is as follows.

1. At 208Y/120 V, three-phase:
   - CPU $208 \times 22 \times 2 \times 1.73$ =   15.8 kVA
   - CPU $208 \times 17 \times 1.73$ =   6.1
   - Disk drive $208 \times 9.5 \times 4 \times 1.73$ =   13.7

   Total, three-phase   35.6 kVA

2. At 120 V, one-phase:
   - Printer $120 \times 11 \times 4$ =   5.3 kVA
   - Printer $120 \times 3 \times 2$ =   0.7
   - Plotter $120 \times 12 \times 6$ =   8.6
   - Terminal $120 \times 4 \times 12$ =   5.8

   Total, one-phase   20.4 kVA

Assuming that the one-phase load is balanced among the three phases of the UPS, the total requirement is

   - Total three-phase load   35.6 kVA
   - Total one-phase load   20.4
   - Load growth, 50 percent   28.0
     Total   84.0 kVA

Select a 100-kVA UPS module rated for 208Y/120-V 60-Hz output.

Example 6.2   Select a battery for the UPS module of Example 6.1. Assume that the load is at 0.8 PF, the load must be delivered for 5 min, the inverter efficiency is 0.95, the battery voltage is 300 V, and the end-of-discharge cell voltage is 1.75 V/cell.

solution   The power to be delivered by the battery is

$$\text{Battery power} = 84.0 \text{ kVA} \times 0.8 \text{ PF} \times 1/0.95$$
$$= 70.7 \text{ kW}$$

The ampere-hour requirement at 5-min discharge rate is

$$\text{Ampere-hours} = 70.7 \text{ kW} \times 1000 \times 5 \text{ min}/60 \times 1/300 \text{ V}$$
$$= 19.6 \text{ Ah}$$

The number of cells is

$$\text{Cells} = \frac{300 \text{ V}}{1.75 \text{ V}} = 171.4 \ (172 \text{ cells})$$

The battery current at the end of discharge is

$$\text{Battery current} = 70.7 \text{ kW} \times 1000 \times 1/300 \text{ V}$$
$$= 236 \text{ A}$$

Example 6.3 [9]   Each module of a two-module redundant UPS is rated for a maximum input line current of 120 percent, of which 100 percent is available to supply the load and 20 percent is available to charge the batteries. Compare the input line current for loads from 100 percent (one-module rating) down to 25 percent for (1) using all the excess module capacity to charge the batteries and (2) limiting the battery charging to 20 percent of each module's rating. Which method is better?

solution   The two options are shown in the following table:

Output load, %	Total module usage		Battery current limited	
	Battery current, %	Line current, %	Battery current, %	Line current, %
100	140	240	40	140
75	165	240	40	115
50	190	240	40	90
25	215	240	40	65

The battery current limited option is preferable for two reasons: First, under the total module usage option, the battery will recharge quickly at first at 140 percent current but will also reach float voltage quickly, and the battery current will have to be reduced. The overall recharging time will not be reduced that much. Second, under the battery current limited option the reduced line current may produce a lower demand charge for the facility.

Example 6.4   A group of parallel-redundant UPS modules is to supply 1200-kVA load with 10 percent margin with one module out of service. The available modules are 100 kVA for $10,000, 250 kVA for $15,000, and 400 kVA for $25,000. Select the least-cost group of modules.

solution   The required output with one unit out is

$$1200 \text{ kVA} \times 1.1 = 1320 \text{ kVA}$$

The available combinations are as follows:

Module size	Number	Cost
100 kVA	14 + 1	$150,000
250	6 + 1	105,000
400	4 + 1	125,000

Select seven 250-kVA modules, at a total cost of $105,000, for a least-cost system.

6.10 Summary

The static UPS is the heart of practically all emergency/standby systems that must ensure continuous power supply to computers and all other critical loads. The components for UPSs have been well developed for power levels from several hundred watts to several thousand kilowatts in single- and multiple-module systems. Most of the other components, such as power-distribution modules, transfer switches, and engine-generator sets, are peripheral to the UPS modules.

REFERENCES

1. ANSI/NFPA 70-1987, "National Electrical Code®."
2. ANSI/IEEE Std. 446-1987, "IEEE Recommended Practice for Emergency and Standby Power Systems for Industrial and Commercial Applications."
3. S. B. Dewan and A. Straughen, *Power Semiconductor Circuits*, Wiley, New York, 1975.
4. A. E. Fitzgerald, C. Kingsley, and A. Kusko, *Electric Machinery*, 3d ed., McGraw-Hill, New York, 1971.
5. K. Sachs, R. W. Larsen, D. P. Decoster, and S. Plato, "A Low-Impedance Uninterruptible Power Technology for Nonlinear Critical Loads," *IEEE Trans. on Ind. Appl.*, vol. IA-23, no. 5, September/October 1987, pp. 796–803.
6. "Specification for a Single Module Uninterruptible Power System," Emerson Electric Co., Santa Ana, Calif.
7. R. T. Laurie, "Designing and Installing a Power System for a Large Computer Center," *EC&M*, January 1986, pp. 76–83.
8. "Modular Power Distribution Systems for Computer Rooms," ISOREG Corp., Littleton, Mass., 1982.
9. M. James, "Specifications and Requirements for Uninterruptible Power Systems," *Intelec '79*, pp. 133–136.

Chapter

7

Batteries

An electric cell is a device that converts the chemical energy contained in its active materials directly into electric energy by means of an oxidation-reduction electrochemical reaction. A battery is formed by interconnecting one or more cells electrically. Batteries may consist of primary (nonrechargeable) cells or secondary (rechargeable) cells. In this chapter, only secondary batteries, also called storage batteries, will be treated.

Batteries are a source of energy alternative to fossil fuels for emergency/standby systems. They are used to provide power to dc loads directly, as in telephone systems, emergency lighting equipment, and control power for switchgear. They are used, through rotary and static UPSs, to provide power for ac loads or, through standby inverters, to provide power for lighting and industrial processes.

7.1 Codes and Standards

The following standards apply to secondary batteries for emergency/ standby systems:

1. ANSI/IEEE Std. 450-1980, "IEEE Recommended Practice for Maintenance, Testing, and Replacement of Large Lead Storage Batteries for Generating Stations and Substations" [1]

2. ANSI/IEEE Std. 485-1983, "IEEE Recommended Practice for Sizing Large Lead Storage Batteries for Generating Stations and Substations" [2]

3. ANSI/IEEE Std. 484-1981, "IEEE Recommended Practice for Installation Design and Installation of Large Lead Storage Batteries for Generating Stations and Substations" [3]

4. ANSI/IEEE Std. 484A-1984, Supplement to IEEE 484-1981 [4]

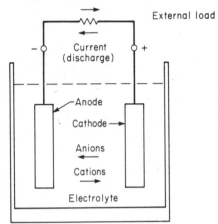

Figure 7.1 Model of a cell under discharge conditions.

5. ANSI/NFPA 70-1987, National Electrical Code®,* Art. 480, "Storage Batteries" [5]

6. ANSI/IEEE Std. 446-1987, "IEEE Recommended Practice for Emergency and Standby Power Systems for Industrial and Commercial Applications," Art. 4.7, "Battery Systems" [6]

7.2 Electrochemistry of Batteries

The model of an electric cell is shown in Fig. 7.1. The actual cell consists of three major components: the anode (the reducing material), the cathode (the oxidizing agent), and the electrolyte which provides the necessary internal ionic conductivity. Electrolytes are usually liquid; some electrolytes are solid but are ionic conductors at room temperature. In practical cells, separators are used between the anode and cathode electrodes; grid structures or materials are added to each electrode to reduce internal resistance and provide terminals. A simplified explanation for the electrochemical reactions will be provided for the lead-acid cell and the nickel-cadmium cell, which make up the bulk of the emergency/standby batteries.

Lead-acid cell

The charged lead-acid cell (of a battery), Fig. 7.1, uses highly reactive sponge lead for the anode (negative electrode), lead dioxide for the

---

*National Electrical Code® and NEC® are registered trademarks of the National Fire Protection Association, Inc., Quincy, Mass.

cathode (positive electrode), and sulfuric acid solution for the electrolyte. As the cell discharges, the active materials at both electrodes are converted into lead sulfate. The sulfuric acid electrolyte produces water. The reactions, based on the double-sulfate theory, are

$$Pb + PbO_2 + 2H_2SO_4 \overset{\text{Discharge}}{\underset{\text{Charge}}{\rightleftharpoons}} 2PbSO_4 + 2H_2O$$

The water (H_2O) on the right-hand side dilutes the electrolyte. The state of charge can be determined by measuring the specific gravity, which decreases on discharge and increases on charge. At the end of the charge, electrolysis of water occurs, producing hydrogen at the negative electrode and oxygen at the positive electrode.

Nickel-cadmium cell

Nickel-cadmium cells are manufactured in two basic types: sealed and vented. Some sealed cells incorporate a high-pressure vent as a safety measure (operation is at 100 to 300 psig). The sealed cell is designed for applications which are lightweight and portable and for long operating life. Nickel-cadmium cells are frequently used as replacements for primary AA, C, and D cells in cylindrical, button, rectangular, and oval form.

In the sealed cell, nickel hydroxide ($NiOOH$) is the active material of the cathode (positive electrode) and cadmium is the active material of the anode (negative electrode). The electrolyte is a potassium hydroxide solution. The electrolyte does not enter into the reaction. Only oxygen ions travel between the electrodes. During discharge, the cadmium of the anode is oxidized to cadmium hydroxide and releases electrons to the external circuit. During discharge, the charged nickel hydroxide ($NiOOH$) goes to a lower valence state [$Ni(OH)_2$] by accepting electrons from the external circuit. During charge, the reactions are reversed. The reactions are the following:

$$Cd + 2H_2O + 2NiOOH \overset{\text{Discharge}}{\underset{\text{Charge}}{\rightleftharpoons}} 22Ni(OH)_2 + Cd(OH)_2$$

The vented cell is designed to provide high-rate discharge service and fast recharging for engine-starting emergency/standby applications. It is generally smaller and lighter and requires less maintenance than lead-acid cells. During overcharge, the resealable vent allows hydrogen and oxygen gases generated by the electrolysis of water to escape to the atmosphere. The vent isolates the electrolyte so that it cannot pick up carbon dioxide from the atmosphere and become carbonated. As a result, deionized water may have to be added at regular intervals.

7.3 Types of Batteries

Two types of batteries are used for emergency/standby service: (1) lead-acid and (2) nickel-cadmium. The nickel-cadmium battery is more expensive than the lead-acid type, but it generally is more rugged, requires less maintenance, and exhibits a longer life. The sealed lead-acid battery is displacing the nickel-cadmium battery, however.

Lead-acid batteries

Technological advances in lead-acid batteries in recent years have resulted in maintenance-free batteries for emergency/standby systems [7]. Three types are available: (1) the conventional flooded battery, (2) the sealed gelled battery, and (3) the sealed liquid-immobilized battery. The amount (or concentration) and distribution of the sulfuric acid electrolyte within the battery determine the battery capacity. The three types of lead-acid batteries can be contrasted by the nature of the acid. The flooded battery, which has been the convention in the industry, has liquid acid covering the entire electrode stack. Because the liquid acid is mobile, acid outside the plates, i.e., that above and below the plates in addition to that between the plates, is available for reaction. The acid in a gelled battery is immobilized between the plates by addition of silicon to the sulfuric acid. The acid in an immobilized battery is retained in an absorbent separator between the plates. In both the gelled and liquid-immobilized batteries, the acid available for reaction is only that in and between the plates.

At short-time high-discharge currents, the capacities delivered by the three battery types are identical. The acid being utilized resides in and adjacent to the plates. Any acid outside the plate stack does not have time to diffuse to the plate surfaces. As the discharge time is increased and the current is reduced, the capacity of the flooded battery becomes greater than that of the equivalent gelled and liquid-immobilized types. The acid outside the plate stack has time to enter the plate surface for the reaction.

Generally, flooded batteries of the same weight and volume as gelled or immobilized batteries will deliver more capacity because of the mobility of the acid. However, the flooded battery will usually have an excess amount of electrolyte above the plates to minimize the frequency of watering. The gelled or immobilized batteries can utilize this space for extra plate height to increase the capacity of the battery at the expense of adding to the weight.

Watering is necessary for a flooded battery. Water loss is the result of evaporation and electrolysis of water that occurs in lead-acid batteries, particularly when the batteries are overcharging. When the battery is near its full state of charge, only a portion of the charging current replenishes the charge on the battery plates. The remaining current electrolyzes water to release hydrogen and oxygen through the battery vents. Since gelled and liquid-immobilized batteries allow recombination of the oxygen that is formed during charging, water loss from these batteries is extremely low compared to that from flooded batteries.

Venting with proper room air circulation is desirable for the safe operation of all batteries. In flooded cells, the evolved hydrogen and oxygen gases pass through the vent caps. Gases will be emitted from gel and liquid-immobilized batteries only when their vent pressures are exceeded, e.g., 0.5 to 8 psi. Most of the emitted gas is hydrogen, indicating efficient oxygen recombination. Significant gas release occurs in the gel cell, with a 0.5-psi vent, only when the cell voltage exceeds 2.35 V, because of a malfunction of the charger, or high battery temperature.

The lead in battery grids is combined with either antimony or calcium. The amount of antimony ranges from 5 to 12 percent. It is used to strengthen the lead grid and to facilitate manufacturing without increasing the internal resistance of the battery. As an alternative, the grids are hardened with up to 0.1 percent calcium. The result is a battery of lessened self-discharge, increased life, and floating-charge requirement reduced to about one-fifth of the usual value. Examples of lead-acid batteries are shown in Figs. 7.2 to 7.4.

Figure 7.2 Flooded electrolyte lead-acid battery. (*Courtesy GNB–Industrial Battery Co.*)

Figure 7.3 Sealed lead-acid battery using glass-fiber separator. (*Courtesy GNB–Industrial Battery Co.*) [9].

Gelled electrolyte sealed cells are made by using flat-plate construction as in flooded electrolyte cells [8]. For large long-life batteries for float service, tubular positive plates are employed. These tubular plates are made by using lead-calcium alloy to obtain the lowered maintenance and improved voltage control of this alloy while main-

Figure 7.4 Lead-acid battery bank for UPS. (*Courtesy Emerson Electric Co.*)

taining the longer cycle life and improved active material utilization of the tubular plate.

The absorbed electrolyte sealed battery has a low internal resistance, limited electrolyte for reaction, and limited capacity for deep discharge. Suitable applications include UPS where discharge times of 5 to 15 min are typical. The gelled electrolyte sealed battery has a higher internal resistance and slightly lessened discharge rate capability at very high rates of discharge. The battery is suitable for long-life, up to 4-h discharge time, applications.

Nickel-Cadmium Batteries [10]

The cells for nickel-cadmium batteries are first classified into sealed and vented cells, as described in Sec. 7.2, and then into pocket plates and sintered plates.

In the pocket plate cell, the positive and negative plates are usually similar in construction; they consist of perforated pockets which contain the active materials. The pockets in the positive plates are filled with nickel hydroxide, to which is added graphite to increase the conductivity. The pockets of the negative plates are filled initially with cadmium oxide, or cadmium hydroxide, which is reduced to metallic cadmium on first charge. Iron is usually added to the cadmium in order to maintain the required degree of fineness of the cadmium. The pockets are pressed into grids, usually made of nickel-plated steel. The plates are assembled into elements: separators are placed between the plates of opposite polarity and to the inside of the steel container of the cell. Terminal posts are sealed through glands. Vent caps are spring-loaded. A pocket plate vented cell is shown in Fig. 7.5.

In the sintered plate cell, the plates consist of a highly porous structure of nickel impregnated with the active materials nickel oxide and cadmium. The plaques are the sintered metal bodies before being impregnated. After that they are called plates. The plaque is made by sintering carbonyl nickel in a woven nickel wire cloth grid.

A sealed cylindrical cell uses a cylindrical nickel-plated steel case as a negative terminal and a cell cover as the positive terminal. The sintered plates, which are wound to form a compact roll, are isolated from each other by a porous separator. An insulated seal ring separates the positive cover from the negative can.

A vented rectangular cell, as shown in Fig. 7.6, consists of flat positive and negative nickel-cadmium plates separated by materials which act as a gas barrier and separator. The separator is a continuous, thin, porous multilaminate of woven nylon and cellophane. The cell container, usually made of nylon, consists of a cell jar and matching cover that are permanently joined at assembly to provide a sealed enclosure that prevents electrolyte leakage or contamination.

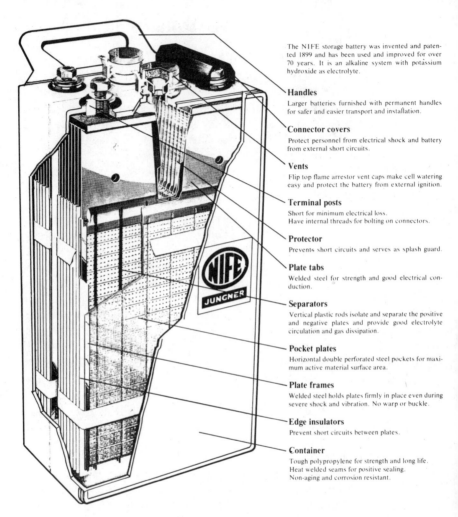

The NIFE storage battery was invented and patented 1899 and has been used and improved for over 70 years. It is an alkaline system with potassium hydroxide as electrolyte.

Handles
Larger batteries furnished with permanent handles for safer and easier transport and installation.

Connector covers
Protect personnel from electrical shock and battery from external short circuits.

Vents
Flip top flame arrestor vent caps make cell watering easy and protect the battery from external ignition.

Terminal posts
Short for minimum electrical loss. Have internal threads for bolting on connectors.

Protector
Prevents short circuits and serves as splash guard.

Plate tabs
Welded steel for strength and good electrical conduction.

Separators
Vertical plastic rods isolate and separate the positive and negative plates and provide good electrolyte circulation and gas dissipation.

Pocket plates
Horizontal double perforated steel pockets for maximum active material surface area.

Plate frames
Welded steel holds plates firmly in place even during severe shock and vibration. No warp or buckle.

Edge insulators
Prevent short circuits between plates.

Container
Tough polypropylene for strength and long life. Heat welded seams for positive sealing. Non-aging and corrosion resistant.

Figure 7.5 Pocket plate sintered nickel-cadmium cell with plastic container. (*Courtesy SAB NIFE INC.*)

The vent cap assembly in the cover releases excessive gases during charging. Except when releasing gas, the vent automatically seals the cell to prevent electrolyte spillage and entry of oxygen or foreign material into the cell.

7.4 Characteristic Curves

Battery characteristics, that is, behavior during discharge and charge, are a function of the cell materials, charge or discharge rates, temperature, electrolyte, and physical dimensions. Manufacturers provide

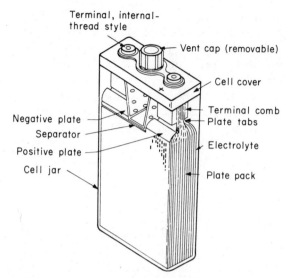

Figure 7.6 Sintered-plate-type vented nickel-cadmium cell.

these characteristics, usually for standardized conditions. Examples of characteristics for lead-acid and nickel-cadmium batteries will be given in the following subsections.

Lead-acid battery

The characteristic discharge and charge curve is shown in Fig. 7.7. The nominal voltage of the cell is 2.0 V. The voltage on open circuit ranges from 2.12 V for 1.28 specific gravity (SG) to 2.05 V for 1.21 SG.

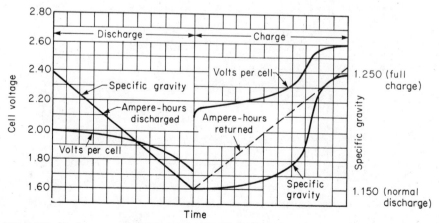

Figure 7.7 Performance characteristics of lead-acid batteries [11].

The discharged end voltage is shown as 1.75 V. The nominal discharge time for stationary batteries is 8 h, and that for starting and lighting batteries is 20 h, at 25°C. A specific gravity of 1.26 to 1.28 is used for fully charged starting batteries and down to 1.21 for stationary standby batteries. The specific gravity decreases by 0.125 down to 1.150 from a fully charged to a fully discharged condition. Note in Fig. 7.7 that the linear decline of the ampere-hours matches the specific gravity, which is therefore an accurate measure of the discharge level of the battery. Discharge curves at rates from 1 to 12 h to the same specific gravity are shown in Fig. 7.8.

Nickel-cadmium battery [10]

Typical charge-discharge characteristics of a vented cell are shown in Fig. 7.9. The nominal cell voltage is 1.2 V. The specific gravity of the electrolyte does not change during the processes; it cannot be used as a measure of state of charge. Nickel-cadmium battery capacity is usually given at the 5-h rate. The difference between sealed and vented batteries during charging is shown in Fig. 7.10. The voltage of the sealed cell remains almost constant. The effect of discharge rate of a vented cell from 1.0 C (nominal rate) to 7.0 C is shown in Fig. 7.11.

7.5   Sizing

The standards for sizing storage batteries for emergency/standby service are listed in Sec. 7.1 and are also described by Migliaro [12]. Battery manufacturers' bulletins provide data on a per cell basis and describe the sizing procedure. UPS manufacturers provide data on the

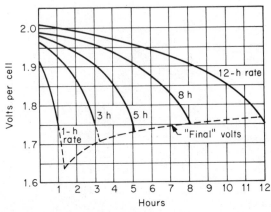

Figure 7.8 Discharge curves of lead-acid batteries at different hour rates [11].

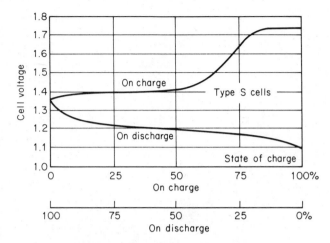

Figure 7.9 Relation of voltage to state of charge of nickel-cadmium cells. A 7-h charge current equals the normal 5-h discharge current [10].

required number of cells by battery manufacturer, rating of their UPSs, and protection time for the critical load.

The stationary battery in standby service is usually operated in "full-float operation," which is defined as a dc system with the battery, battery charger, and load all connected in parallel with the battery charger supplying the normal dc load. The battery will deliver current only when the load exceeds the charger output. The cells are not designed for frequent charge-discharge cycles, typically only two full discharges per year.

The cells are arranged with n positive plates and $n + 1$ negative plates to ensure that all of the positive plate active material is utilized. The assembly of the plates, separators, connectors, and posts is called an element. The element is placed in a container that holds the electrolyte.

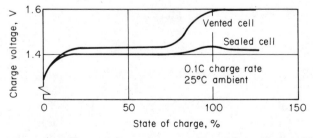

Figure 7.10 Charge voltage characteristics: sealed and vented cells.

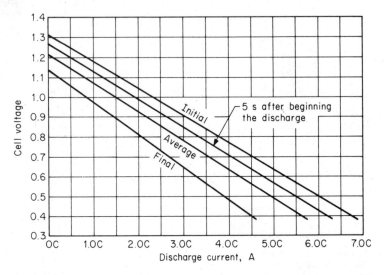

Figure 7.11 High-rate discharge characteristics of nickel-cadmium cells, type S [10].

The nominal ratings of lead-acid cells are at the 8-h rate, 77°F, 1.215 SG, to 1.75 volts per cell (V pc) (discharged). Nickel-cadmium values are based on the 8-h rate, 77°F, to 1.14 V pc, unless otherwise noted. The specific sizing process requires determining the number of cells, and the cell capacity, expressed as ampere-hours (Ah) or the number of positive plates of a particular cell.

Number of cells

A lead-acid cell has a nominal voltage of 2.0 V pc. However, the typical charging voltage is 2.33 V pc and the end discharge voltage is 1.75 V pc. Either the number of cells can be selected for the nominal dc voltage and the end points calculated or the end points can be specified and the number of cells calculated. For example, for a nominal 120-V dc system, 60 cells are required. The end-point voltages are:

$$\text{Maximum voltage} = 60 \times 2.33 = 140 \text{ V}$$

$$\text{Minimum voltage} = 60 \times 1.75 = 105 \text{ V}$$

A nickel-cadmium cell has a nominal voltage of 1.2 V pc. The typical charging voltage is 1.52 V pc, and the end discharge voltage is 1.14 V pc. For the same nominal 120-V dc system, 100 cells are required. The end-point voltages are:

$$\text{Maximum voltage} = 100 \times 1.52 = 152 \text{ V}$$
$$\text{Minimum voltage} = 100 \times 1.14 = 114 \text{ V}$$

The minimum voltage corresponding to the end point of discharge is critical to permit operation of dc contactors and relays. These are typically rated to operate to -20 percent of nominal voltage. For example, a relay rated at 125 V dc will operate down to $0.8 \times 125 = 100$ V. The relay would operate at the minimum voltage in both of the battery examples given above.

Sizing

The duty cycle is defined as the load currents a battery is expected to supply for specified time periods. The loads on a stationary battery are categorized as continuous or noncontinuous. Continuous loads are those supplied throughout the duty cycle. Noncontinuous loads are those supplied for only a portion of the duty cycle; those lasting 1 min or less are termed "momentary loads"; those that can occur at any time are termed "random loads." For lead-acid cells, loads lasting less than 1 min are considered to last for the full 1 min; discharge rates are given only down to the 1 min. When a discrete duty cycle for a period of 1 min can be defined, the momentary load is taken as the maximum current during the 1-min period. When the discrete duty cycle cannot be defined, the momentary load is taken as the sum of all currents during the 1-min period. Nickel-cadmium cells have published discharge rates down to 1 s.

Batteries for emergency/standby service for a duty cycle are sized by using the positive-plate or ampere-hour (Ah) method. Batteries for UPSs for a constant power drain are sized by using the kilowatt method. The following factors are provided by the battery manufacturer for each cell size (plate area):

K_T Used for sizing based on the Ah method. Capacity rating factor at the T-minute discharge rate at 77°F to a definite end-of-discharge voltage.

R_T Used for sizing based on the positive-plate method. Capacity rating factor at the T-minute discharge rate at 77°F to a definite end-of-discharge voltage.

The following correction factors are provided by the standards:

k_a Aging factor, usually 1.25.

k_d Design margin factor, usually 1.15.

k_t Temperature correction factor, usually 1.0 for temperatures above 77°F; see manufacturers' data for temperatures below 77°F.

The sizing of a lead-acid cell will be described by example in Sec. 7.7.

For UPS battery sizing, battery manufacturers can supply kilowatt rates for their cells at 77°F, a specific charge SG, and to various discharge end voltages. These rates are given as F_W, kilowatts per cell. The kilowatt method is described by example in Sec. 7.7.

7.6 Maintenance

All stationary batteries for full-float operation in an emergency/standby system require maintenance [13]. A properly maintained battery will allow the user to realize optimum battery life. The factors in battery maintenance include the following:

Standards

ANSI/IEEE Std. 450-1980 [1] includes recommended practice for maintenance, testing, and replacement of large lead-acid batteries. A standard for nickel-cadmium batteries is in preparation. United States standards of reference for the capacity of lead-acid cells is based upon the 8-h rate at 77°F, 1.215 SG, discharged to 1.75 V pc. For nickel-cadmium batteries, the capacity is based upon the 8-h rate at 77°F discharged to 1.14 V pc.

Maintenance program

The maintenance program should be consistent and regular; it must address the following:

- Designed for the specific battery
- Collected data corrected to standard reference for comparisons
- Visual checks
- Proper procedures for measurements and tests
- Cleaning of jars, covers, terminal posts, connectors, and flame arrester caps
- Pilot cells rotated quarterly

Visual inspections

Visual inspections are recommended monthly (general) and yearly (detailed). The batteries should be inspected for the following:

- Cracks and structural damage of jars
- Seals

- Color of plates
- Sulfation
- Electrolyte level
- Corrosion
- Cables
- Gassing
- Freezing
- Hydration
- Mossing
- Sediment

Temperature

Electrolyte temperature should be read and recorded when SG or voltage readings are taken. These readings are used to correct SG and voltage to standard readings. The differential temperature between cells must not exceed 5°F. Optimum battery performance is obtained when electrolyte temperature is maintained at 77°F.

Specific gravity

Specific gravity should be read and recorded at least monthly on the pilot cells and quarterly on all of the cells. For lead-acid cells, specific gravity is a good indication of charge. Corrections for electrolyte temperature and level must be applied to adjust the SG readings to a standard reference. Corrected values should be compared with previous data. Batteries may have different nominal specific gravities. New high-performance cells for UPS service may have nominal specific gravities as high as 1.30.

Voltage

Voltages should be read and recorded at least monthly on the pilot cells and quarterly on all of the cells. Correction for electrolyte temperature should be made. Readings should be compared with previous data. Open-circuit voltage is approximated by SG + 0.84 V pc. Float voltage is related to cell type, plate alloy, and cell SG. Too high a float voltage will result in overcharging and reduced battery life. In some installations, slightly higher float voltage is selected to reduce or eliminate the need for periodic equalizing charges.

Other considerations

- Connection resistance
- Capacity tests
- Individual charging
- Jumper cells
- Battery racks

Nickel-cadmium batteries

All of the procedures and tests for lead-acid batteries are valid for nickel-cadmium batteries except for specific gravity. The nickel-cadmium cell electrolyte is a solution of potassium hydroxide in water with between 1.180 and 1.200 SG. The electrolyte does not enter into the reaction and is therefore not an indication of state of charge.

Example 7.1 [12]   Assume that a cell must be selected for the duty cycle of Fig. 7.12a by using the positive-plate method. Specify the combined correction factor as $K_t k_d k_a = 1.11 \times 1.15 \times 1.25 = 1.6$. The capacity rating factors R_T for end voltages of 1.75 V pc and 1.81 V pc are given by Table 7.1. Use 1.75 V pc for this calculation.

solution

1. Section 1 (Fig. 7.12b).   At 1100 A for 1 min, $R_T = 197$ A/pos. plate; then

$$F_s \text{ (uncorr. cell no.)} = 1100/197 = 5.58 \text{ pos. plates}$$
$$F \text{ (corr.)} = 1.6 \times 5.58 = 8.93 \text{ pos. plates}$$

2. Section 2 (Fig. 7.12c).   At 1100 A for 120 min, $R_T = 57.3$ A/pos. plate; and at 200 A for 119 min, $R_T = 57.5$ A/pos. plate. Then

$$F_s = (1100/57.3) + (-900/57.5) = 3.55 \text{ pos. plates}$$
$$F = 1.6 \times 3.55 = 5.68 \text{ pos. plates}$$

3. Section 3 (Fig. 7.12a).   At 1100 A for 180 min, $R_T = 44$ A/pos. plate; at 200 A for 179 min, $R_T = 44.2$ A/pos. plate; and at 50 A for 60 min, $R_T = 89$ A/pos. plate. Then

$$F_s = (1100/44) + (-900/44.2) + (-150/89) = 2.95 \text{ pos. plates}$$
$$F = 1.6 \times 2.95 = 4.72 \text{ pos. plates}$$

TABLE 7.1   Capacity Rating Factor for Sample Cell

Discharge period, min	R_T, A/pos. plate, to 1.75 V pc	R_T, A/pos. plate, to 1.81 V pc
1	197	143
60	89	75
119	57.5	52.1
120	57.3	52
179	44.2	40.9
180	44	40.7

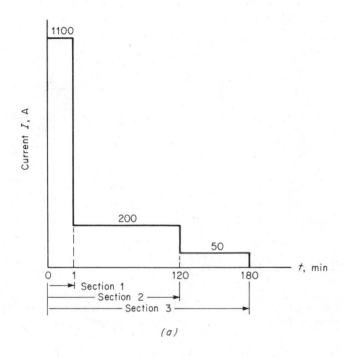

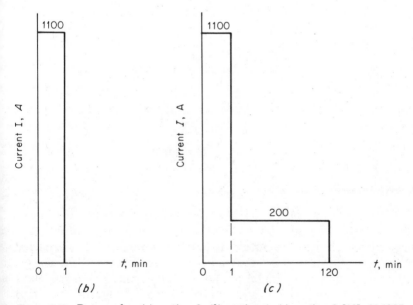

Figure 7.12 Duty cycles: (a) section 3, (b) section 1, (c) section 2 [12]. (©*1985 IEEE*)

TABLE 7.2   Constant-Wattage Discharge Ratings UPS 80, 1- to 60-min Rate, kW/cell at 77°F [14]

Cutoff voltage, volts/cells	Power per cell, kW, for the following operating times to cutoff voltage, in minutes										
	1	2.5	5	7.5	10	15	20	30	40	50	60
1.85	.267	.242	.212	.192	.178	.157	.140	.115	.097	.085	.078
1.80	.367	.317	.263	.230	.207	.175	.155	.125	.105	.092	.082
1.75	.442	.387	.295	.263	.235	.190	.165	.130	.110	.095	.083
1.70	.483	.430	.350	.293	.255	.205	.175	.135	.113	.098	.084
1.65	.525	.470	.372	.308	.268	.213	.180	.139	.117	.101	.085
1.55	.600	.517	.413	.337	.283	.222	.185	.143	.120	.103	.086

Note: Although figures express kilowatts per cell, cells must be used in multiples of six (6) cells.

Section 1 requires the largest number of positive plates, 8.93 rounded to 9, and the total number of plates in the cell is thus $2 \times 9 + 1 = 19$ plates per cell.

Example 7.2   A UPS must deliver 20 kW for 20 min from its battery system. The efficiency is assumed to be 0.95. The battery system consists of 120 cells with the constant-wattage discharge ratings shown in Table 7.2. What will be the end-of-discharge cell and battery system voltage?

solution   The power to be delivered by the battery system is

$$P_b = 20,000/0.95 = 21,053 \text{ W}$$

The power per cell is

$$P_c = 21,053/120 = 175.4 \text{ W, } 0.175 \text{ kW}$$

From Table 7.2, at 0.175 kW and 20 min time, the cutoff cell voltage is 1.70 V pc. The end-of-discharge battery system voltage is

$$V_b = 120 \times 1.70 = 204 \text{ V}$$

7.7   Summary

Batteries for emergency/standby systems are assembled from lead-acid or nickel-cadmium cells to meet the requirements for voltage, discharge current, duration of discharge, and other factors. They are used for emergency lighting, direct supply of dc equipment, UPS power, and engine cranking. Standards and extensive manufacturers' data are available to assist the designer in selecting the required type and combination of cells for the application.

REFERENCES

1. ANSI/IEEE Std. 450-1980, "IEEE Recommended Practice for Maintenance, Testing, and Replacement of Large Lead Storage Batteries for Generating Stations and Substations."
2. ANSI/IEEE Std. 485-1983, "IEEE Recommended Practice for Sizing Large Lead Storage Batteries for Generating Stations and Substations."
3. ANSI/IEEE Std. 484-1981, "IEEE Recommended Practice for Installation Design and Installation of Large Lead Storage Batteries for Generating Stations and Substations."
4. ANSI/IEEE Std. 484A-1984, Supplement to IEEE 484-1981.
5. ANSI/NFPA 70-1987, "National Electrical Code®," Art. 480, "Storage Batteries."
6. ANSI/IEEE Std. 446-1987, "IEEE Recommended Practice for Emergency and Standby Power Systems for Industrial and Commercial Applications," Art. 4.7, "Battery Systems."
7. B. L. McKinney, T. J. Dougherty, and M. Geibl, "The Comparison of Flooded, Gelled and Immobilized Lead-Acid Batteries," *Intelec '84,* pp. 41–44.
8. J. J. Kelley and C. K. McManus, "Sealed Lead Acid Batteries," *Intelec '86,* pp. 43–47.
9. J. B. Doe and P. W. Lemke, "Separators and Their Effect on Lead-Acid Battery Performance," *Intelec '86,* pp. 67–71.
10. G. W. Vinal, *Storage Batteries,* 4th ed., Wiley, New York, 1955.
11. D. G. Fink and H. W. Beaty, *Standard Handbook for Electrical Engineers,* 12th ed., McGraw-Hill, New York, 1987.
12. M. W. Migliaro, "Considerations for Selecting and Sizing Batteries," *Conf. Record, IEEE-IAS-1985 Annual Meeting,* pp. 345–353.
13. M. W. Migliaro, "Maintaining Stationary Batteries," *Conf. Record, IEEE-IAS-1986 Annual Meeting,* pp. 1011–1017.
14. "UPS 80 Sealed Lead Acid Battery Bulletin," Globe Battery Division, Johnson Controls, Inc., Milwaukee, Wis., 1983.

8

Power Distribution Units

Emergency power systems which supply lighting equipment, elevators, fire pumps, and the like are wired with conventional conduits, panels, circuit breakers, and wire. Data processing equipment, e.g., CPUs, disk and tape drives, terminals, and printers, have different power requirements. The voltage quality must be high; both 60- and 415-Hz power are needed; and the arrangement of the equipment may be changed frequently. Power distribution units which stand on the computer room floor to distribute power to surrounding equipment have been developed. These units accept power from the utility source, or from a UPS, monitor it, transform and regulate the voltage, and distribute the power through circuit-breaker-protected feeders to the data processing equipment.

8.1 Components of the Power Distribution Unit

A power distribution unit feeding computer loads is illustrated in Fig. 8.1 [1]. The principal functions of the unit are the following:

1. Transform the input voltage, usually 480 V, three phase, to 208Y/120-V power for the three- and one-phase computer loads.

2. Regulate the output load voltage against the effects of input voltage fluctuations, load switching, and transients. This function is not required for power distribution units supplied from a UPS.

3. Isolate the loads from electrical noise on the input line.

4. Provide circuit breaker protection for the equipment on the computer room floor.

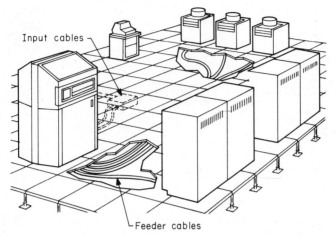

Figure 8.1   Illustration of power distribution unit on computer room floor.

These functions will be described in the following sections.

8.2   Power Distribution Unit

A power distribution unit serves one of the following functions:

1. *Power distribution only,* includes terminals for input supply from a building feeder, UPS, or dedicated transformer, terminals and cables for circuits and circuit breakers to supply the load equipment, and meters, monitor, and/or alarm panel.

2. *Power distribution plus isolation,* includes components of item 1 plus a shielded isolation transformer.

3. *Power distribution plus isolation and voltage regulation,* includes components of item 1 plus a manual or automatic voltage-regulating transformer.

A commercial power distribution unit is shown in Fig. 8.2 [1]. The pertinent parts, keyed to the numbers in the figure, are the following:

A   Flexible conduits or shielded cables with terminations suitable for the specific loads. The lengths are cut for the initial and future locations of the load equipment.

B   Convenience receptacles to plug in additional equipment. Each receptacle is protected by a circuit breaker.

C   Circuit breakers for the cable circuits and the receptacles.

D   Power monitor cabinet.

E   Rotary tap switches, one for each phase, for manually adjusting the output voltage.

Figure 8.2 Commercial power distribution unit [1]. (*Courtesy ISOREG Corp.*)

F   Shielded isolation transformer (inside cabinet) wound for transformation ratio and taps for the tap switches.

G   Input connection box for three-phase conductors and green wire grounding conductor.

Power distribution units are commonly built with transformers and output ratings up to 225 kVA. A typical unit provides output circuits at three-phase, 208-V, three-pole; one-phase, 208-V, two-pole; and one-phase, 120-V, one-pole levels.

8.3   Voltage-Regulating Transformers

Voltage-regulating transformers are categorized by (1) the range of input voltage for which they deliver regulated output voltage and (2) the response time of the output voltage for sudden changes of input voltage or output load. Tap-changer and ferroresonant transformers are representative types.

Tap-changer transformers

The simplest transformer for transforming and adjusting secondary output voltage is one with taps on either the primary or secondary winding and a manual tap switch. As the tap switch is rotated, the turns ratio of the transformer is changed so that the output voltage will be readjusted to a relatively constant value. Such a transformer is suitable where the input line voltage and the load change infrequently.

Where the output voltage must be corrected frequently, an automatic tap-changer-type transformer using solid-state switches is used as shown in Fig. 8.3. Three primary taps are shown; more can be used to obtain smaller voltage steps. The control circuit, not shown, senses the output voltage; it continuously fires the appropriate thyristor switches to maintain that voltage constant. The thyristors may also be phase-controlled to supplement the regulation between steps. The transformer can be built with shields between the primary and secondary windings to provide full isolation.

Ferroresonant transformers

The ferroresonant-type regulating transformer shown in Fig. 8.4 uses the magnetic saturation characteristic of its magnetic core to maintain constant output voltage [2]. The primary portion of the transformer core is separated physically from the secondary portion by a magnetic leakage shunt, which electrically decouples the two windings. The magnetic flux in the primary portion of the core is proportional to the input line voltage; the magnetic flux in the secondary portion of the core is held at saturation level by the ferroresonating action of the core with the capacitor.

When the input line voltage varies within its range, the amplitude of the primary flux also varies, but the secondary flux remains prac-

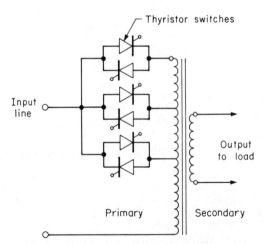

Figure 8.3 Voltage-regulating transformer using solid-state tap switches.

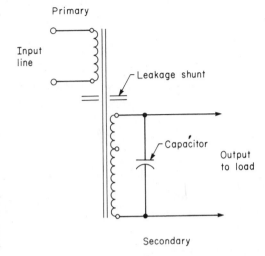

Primary

Input
line

Leakage shunt

Capacitor

Output
to load

Secondary

Figure   8.4   Voltage-regulating
transformer, ferroresonant type.

tically constant. The voltage remains nearly constant, as well, as
shown in the curves of Fig. 8.5 [3]. The transformer also exhibits a flywheel or resonant-circuit effect for short-time discontinuities in the
input voltage; the output voltage remains constant for input voltage
dips to zero for up to one-half cycle.

Ferroresonant transformers are available up to 50 kVA, one-phase
and 300 kVA, three-phase. The ferroresonant transformer can be built
with an electrostatic shield between the primary and secondary windings for full isolation. It has the advantage of regulating without moving parts or auxiliary electronic circuits.

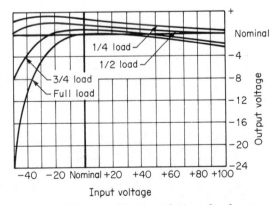

Figure 8.5   Output voltage regulation of a ferroresonant transformer [1]. Percent change of output voltage,
as a function of input voltage, from nominal volume.

8.4   Isolation Transformer

An isolation transformer provides the following four functions.

1. Voltage transformation when required, e.g., 480 to 120-240 V.
2. Insulation against high voltage, where the primary winding may be at a high potential above the level of the secondary winding or above ground.
3. Provision of a separate ground for the secondary circuit.
4. Attenuation of voltage transients from the input line. The shield between the primary and secondary windings serves to protect load equipment from transients that damage semiconductor devices and other components and/or disrupt circuit operation.

A drawing of a shielded isolation transformer is shown in Fig. 8.6 [1]. Each winding is covered by a one-turn metallic shield; a shield is also placed between the windings. Obviously, the one-turn shields are left "open" so as not to short-circuit the transformer.

The connection of a shielded isolation transformer between the input and output cables is shown in Fig. 8.7. On the input side, the black hot wire and the white neutral wire are connected to the primary winding terminals. The shield, the green grounding wire on the input side, the green grounding wire on the output side, and the white neutral wire on the output side are all connected to a common ground point on the case. A grounding wire is then connected from the common point to a local ground. The black, white, and green wires proceed to the load as shown.

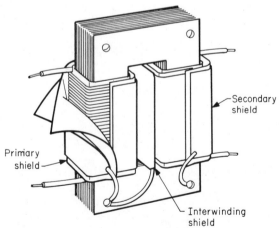

Figure 8.6   Shielded isolation transformer [1].

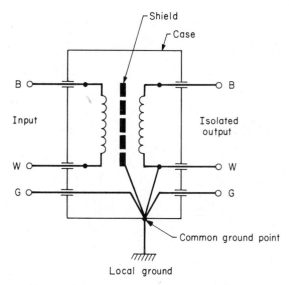

Figure 8.7 Connection of isolation transformer to input line and isolated output line.

The purpose of the shields is to reduce the capacitance between the primary and secondary windings through which transient voltage disturbances reach the load. The equivalent circuit of the transformer is shown in Fig. 8.8. Capacitances C_{11} and C_{22} are the self-capacitances to the core (ground) of the primary and secondary windings. Capacitance C_{12} is that between the windings with no shield in place. When an effective shield is installed, the capacitances of the windings to the shield C'_{12} are shown in parallel with the self-capacitances; the re-

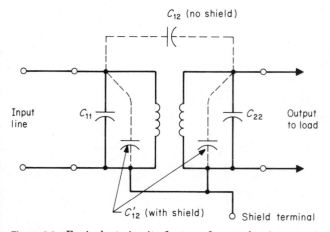

Figure 8.8 Equivalent circuit of a transformer showing capacitance with and without shield.

maining interwinding capacitance C_{12} becomes arbitrarily small. Hence, transient voltages from the input side are shunted to ground by the capacitance C'_{12} to the shield and are not transmitted to the equipment connected to the secondary windings of the transformer.

8.5 Standard Connections

Data processing and communications equipment utilize a variety of one- and three-phase voltage and current power levels. The connections to the equipment are made with plug cords and sockets or hardwired terminal boards or in special junction boxes. NEMA standard plug cord and outlet socket configurations for use in the United States are shown in Fig. 8.9 [3]. The configurations are different outside the United States.

8.6 Summary

Power distribution units serve as a source of regulated, isolated, protected, and independently grounded power for equipment on a computer room floor. The flexible cable feeders extending from the unit

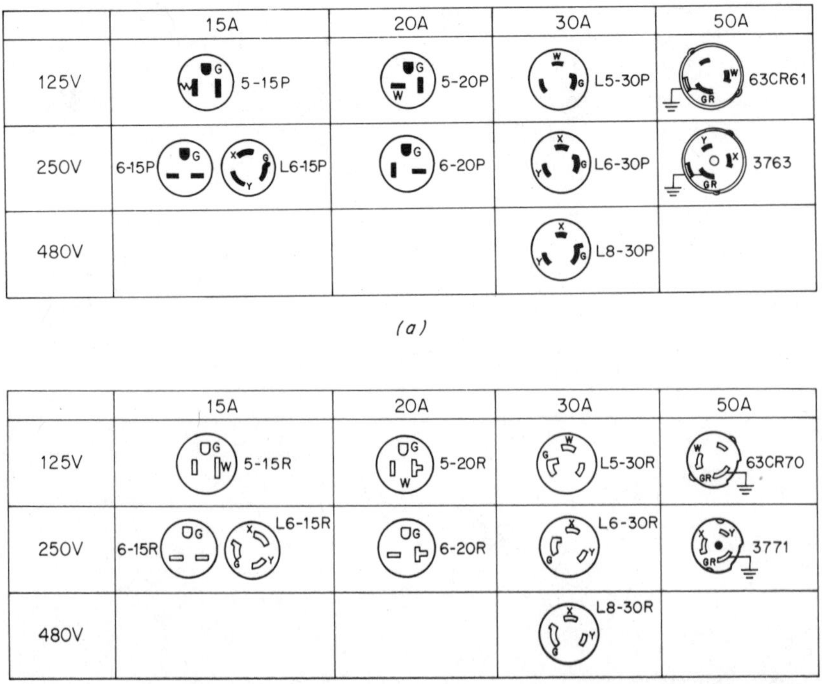

	15A	20A	30A	50A
125V	5-15P	5-20P	L5-30P	63CR61
250V	6-15P / L6-15P	6-20P	L6-30P	3763
480V			L8-30P	

(a)

	15A	20A	30A	50A
125V	5-15R	5-20R	L5-30R	63CR70
250V	6-15R / L6-15R	6-20R	L6-30R	3771
480V			L8-30R	

Figure 8.9   NEMA standard (*a*) plug-cord and (*b*) outlet socket configurations, U.S. [3].

permit equipment to be moved and added without requiring changes in conventional hard-wired conduit distribution systems. The power distribution units can be supplied from utility power, from a UPS, or from either by means of bypass switches. To conserve wire and reduce voltage drop, the units are typically supplied with 480-V three-phase power; the transformer in the unit then steps down the power to 208Y/120 V for three- or one-phase service to the equipment on the computer room floor.

REFERENCES

1. "Modular Power Distribution Systems for Computer Rooms," Isoreg Corp., 1982.
2. A. Kusko and T. Wroblewski, "Core Construction for Large Constant Voltage Transformers," *Intelec '82,* pp. 298–303.
3. "ISOREG Computer Power Conditioners," Isoreg Corp., 1985.

Examples of Emergency/Standby Systems in Use

9

Computer Centers

A computer center includes three levels of equipment:

1. Central processors, tape drives, printers, disk drives, terminals, and other electronic equipment directly involved in data processing

2. Air-handlers, coolant pumps, heat exchangers, lights, and office equipment directly supporting the electronic equipment and the operating personnel

3. Office facilities, lighting, kitchens, general heating and air-conditioning, telephone systems, and other equipment to provide long-term facilities for personnel

All of this equipment requires electric power for its operation. The sensitivity of the equipment to electric supply line disturbances ranges from as little as one-half cycle (8-ms) voltage dip for CPUs to seconds or minutes of voltage dip for air-conditioning equipment that can be restarted when utility or backup engine-generator power is re-established. In planning the center, the equipment is divided into categories by time-voltage sensitivities; matching power supplies must be provided for each. The resultant design of the electric power supply for a computer center is complex not only because of the required levels of power quality but also because of the requirement for bypass and standby facilities to maintain the system.

Since the advent of the computer industry in the early 1960s, the quality of the utility electric supply systems for serving computer centers has not improved [1]. For example, frequent subcycle voltage dips are produced as utilities switch capacitor banks and operate step regulators to control voltage on distribution systems. Transmission and distribution line switching results in further disturbances. In addition to the publicized wide-area blackouts that occur about once per year, more limited blackouts of parts of cities or states occur in the United States about once per week. The owner of a computer center must balance the cost of interruptions in a data processing function with the cost of a UPS to reduce the frequency of the interruptions.

9.1 Terminology

Since UPS and backup power sources for computer centers are not code-required, the terminology is not given in the codes. Commonly used terminology for the computer center loads is as follows:

1. *Critical.* Critical loads require electric power the voltage of which demonstrates less than one-half cycle (8-ms) dip to zero. Such power is provided by static or rotary UPSs from batteries backed up by engine-generator sets.

2. *Essential.* Essential loads require electric power within 10 s after the failure of normal utility electric power. These loads are essentially nonelectronic, i.e., lighting, refrigeration, and ventilation. Such power is provided by engine-generator sets with transfer switches.

3. *Nonessential.* Nonessential loads do not require electric power during the initial failure of normal utility electric power. These loads include nonemergency lighting over a minimum level in nonworking spaces and part of the heating and air-conditioning load. The nonessential load can be divided into priorities; the highest-priority loads can be transferred to the engine-generator sets which have spare capacity.

In addition to the three categories listed above, the loads cited in the NEC®* and NFPA codes must be served. These include egress lights, elevators, and ventilation equipment.

9.2 References

The following references are useful in designing the electrical systems for computer centers:

1. "Guideline on Electric Power for ADP Installations" [2]
2. "Uninterruptible Power Supply Planning Manual" [3]
3. "IBM System/370, Installation Manual, Physical Planning," IBM [4]

---

*National Electrical Code® and NEC® are registered trademarks of the National Fire Protection Association, Inc., Quincy, Mass.

The following standards apply to the electrical systems for computer centers.

4. UL Std. 478, "Electronic Data-Processing Units and Systems" [5]

5. ANSI/NFPA 70-1987, "National Electrical Code®," particularly Art. 100, "Premises Wiring," applied to the flexible conduits from the power distribution centers to the computer equipment, and Sec. 250-26, "Grounding Separately Derived Alternating-Current Systems" [6]

6. ANSI/NFPA 110-1985, "Emergency and Standby Power Systems" [7]

7. ANSI/IEEE Std. 446-1987, "IEEE Recommended Practice for Emergency and Standby Power Systems for Industrial and Commercial Applications" [8]

In addition, manufacturers of UPSs, power distribution units, engine-generator sets, and transfer switches provide extensive technical information on the application of their equipment to electrical systems for computer centers.

9.3 Typical System

Examples of the electrical systems for operating computer centers will be presented later in this chapter. These systems usually include the following parts:

1. *Normal utility.* Supply to the computer center.

2. *60-Hz UPS.* One or more static or rotary UPS modules, depending upon capacity required and degree of redundancy, usually equipped with bypass switches to the normal supply bus.

3. *415-Hz UPS.* One or more static or rotary UPS modules supplied from either the 60-Hz bus or the UPS output bus—usually equipped with line-drop compensators in the 415-Hz feeders.

4. *Standby engine-generators.* One or more sets with diesel-engine- or gas-turbine-driven generators, including voltage regulators, governors, protective relays, synchronizing equipment, and transfer switches.

5. *Power distribution units.* Multiple units on the computer floor to distribute UPS power, including isolation transformers, circuit breakers, distribution panel, and power-monitoring devices.

6. *Mechanical equipment.* For supplying cooling air and water to the

computer equipment, the UPS, and the personnel and including chillers, chilled-water pumps, and air handlers.

A sketch of the electrical distribution system in the computer room itself is shown in Fig. 9.1.

9.4   Example: Affiliated Food Stores [9]

The electrical system shown in Fig. 9.2 supplies two computer systems in a 3300-ft^2 computer center in an office building and a third system in a highly automated warehouse. It is supplied by two 1500-kVA 13.8-kV–208Y/120-V pad-mount transformers, which are fed from the utility's underground network. The power panels 1 and 2 which supply 60-Hz power to the three computer systems are normally supplied by transformer 2 and the single-module 180-kVA UPS. Transformer 1

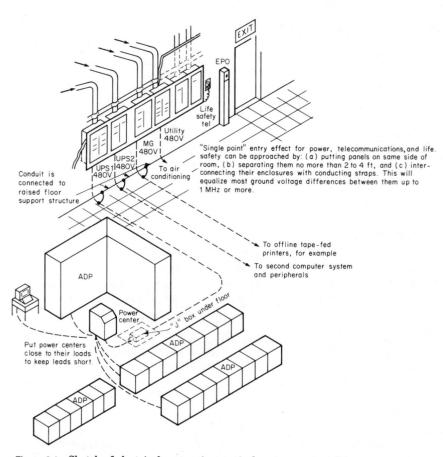

Figure 9.1   Sketch of electrical system in a typical computer room [2].

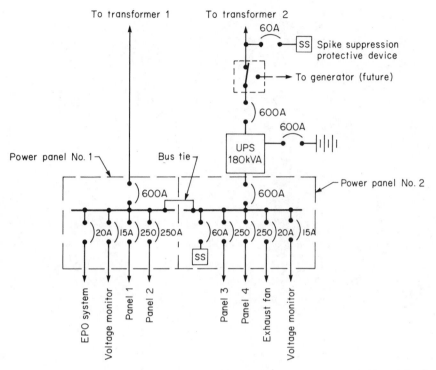

Figure 9.2 Simplified one-line diagram of the electrical system for the Affiliated Food Stores computer facilities [9].

provides alternate and bypass power to the two power panels. A manual transfer switch has been provided for connection to a standby engine-generator set in the future. Selenium-cell-type bidirectional voltage-surge suppressors are connected on the line side of the UPS and on the bus of power panel 2. See Fig. 9.3.

The computer grounding system utilizes driven ground rods around the transformer pads, which are connected to the ground bus in the power panels. Independent insulated grounding conductors are run from the ground bus to orange isolated-ground receptacles in the computer facilities. The electrical boxes and raceways containing the insulated grounding conductor must be grounded by other, independent means.

9.5 Example: Federal Reserve Board [10]

The U.S. Federal Reserve Board relocated and consolidated its computer operations in suburban Washington, D.C. The computer equipment occupies an entire 25,000-ft² floor of the building. The electric power supply to the computers includes a 600-kW 750-kVA 60-Hz

Figure 9.3 Main power panel and 180-kVA UPS, Affiliated Food Stores computer facilities [9].

static UPS with 20 min of battery capacity. No engine-generator backup is provided for extended utility outages; the 20-min period allows orderly shutdown of the computer equipment.

A simplified one-line diagram of the electric supply system to the computer floor is shown in Fig. 9.4. The utility line consists of a 480Y/277-V 3000-A service to a 3000-A switchboard. From the switchboard, one 1600-A feeder is for the UPS and a 1200-A feeder is for the UPS bypass. A second 1200-A feeder supplies the mechanical distribution panel; the panel supplies 13 free-standing air-handling units for the computer units and power to the chilled-water pumps and the lighting circuits. The chillers are supplied directly by 600-A feeders from the switchboard. See Fig. 9.5.

The single static UPS module consists of a 1600-A input circuit breaker, a rectifier/charger, a battery bank, an inverter, a 1200-A output breaker, and a static and electromechanical bypass-switch arrangement. Nonsealed lead-calcium batteries are employed. See Fig. 9.6.

UPS power is distributed from the 1200-A computer distribution panel to seven power distribution units located next to the computer equipment. Each unit consists of a main circuit breaker, a 480-208Y/120-V isolation transformer, and a secondary distribution panel. The feeder cables to the computer equipment are run out of the bottoms of the units in the raised floor below. See Fig. 9.7.

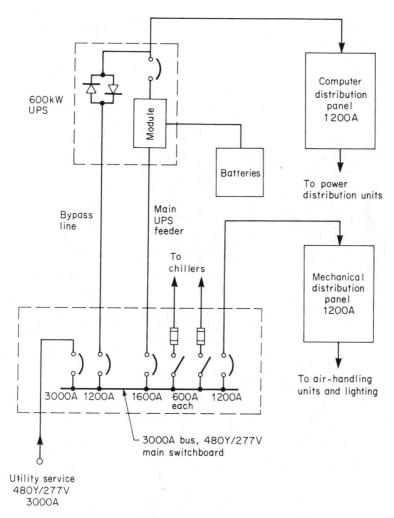

Figure 9.4 Simplified one-line diagram of electrical system for Federal Reserve Board computer center [10].

The computer room grounding system consists of a main copper ground bus in the raised floor interconnecting the ground buses of the power distribution units, the structural members of the raised floor, and a dedicated grounding conductor to the building ground. Both the grounding conductors in the feeder to the units and the secondary neutrals in the units are connected to the ground bus.

9.6 Example: Wakefern Food Corp. [11]

The data center at Wakefern Food Corp., a large cooperative food retailer in New Jersey, handles hundreds of orders, stock inventory,

Figure 9.5   Main 3000-A switchboard, Federal Reserve Board computer center [10].

Figure 9.6   Single-module 600-kW UPS Federal Reserve Board computer center [10].

Figure 9.7 Computer distribution panel, 1200 A, Federal Reserve Board computer center [10].

pricing data, and related information for nearly 200 supermarkets located in five northeastern states. The data center communicates with a minicomputer in each store, obtains orders, processes the orders on two IBM 3083 B-16 mainframes, and sends the information to warehouses where the orders are filled and shipped. The data center also incorporates 500 on-line terminals to aid dispatching, invoicing, accounting, and other paper work.

Figure 9.8 is a one-line diagram of the electric power supply system for the data center. Utility power is delivered at 480Y/277 V to a main switchboard for the entire facility. The switchboard feeds building power directly. The 60- and 415-Hz UPSs, as well as the support mechanical equipment, are fed through two automatic transfer switches. The alternate source is a 1000-kW diesel engine-generator set. See Fig. 9.9.

The UPS consists of one 400-kVA 60-Hz module and two 80-kVA 415-Hz modules. The modules operate from a common lead-acid battery bank. Reliability is enhanced by a solid-state bypass switch on the 60-Hz

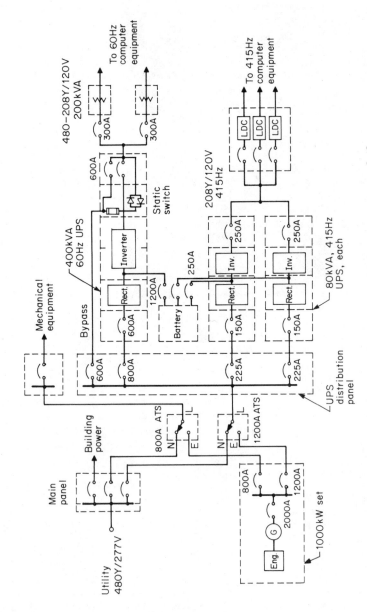

Figure 9.8 One-line diagram of electrical system at Wakefern data processing center [11].

Figure 9.9   Main switchboard and transfer switches, Wakefern data processing center [11].

module and the redundant 415-Hz modules. The 415-Hz feeders are equipped with line-drop compensators (LDCs). See Fig. 9.10.

UPS power at 60 Hz is distributed in the computer area from two 200-kVA 480-208Y/120-V power distribution units. Each unit is equipped with a shielded isolation transformer, a distribution panel with 168 circuit breaker spaces, and flexible multiconductor cables of proper size and length with appropriate connectors for the equipment supplied.

9.7   Example: Amoco Computer Center [12]

Amoco Products Co., in relocating its headquarters to Houston, Texas, required a clean continuous power network to support a 60,000-ft^2 data processing center capable of accommodating computers and staff to the year 1990.

A one-line block diagram of the electric supply system is shown in Fig. 9.11. The utility 4160-V service enters the customer-owned unit substation, from which 480Y/277-V power is distributed to the system. The alternate source consists of three 100-kW diesel engine-generator sets. One set of automatic transfer switches serves the 60-Hz UPS load. The second set of switches serves the emergency system, including

Figure 9.10   UPS room, 60- and 415-Hz modules, Wakefern data processing center [11].

emergency lighting panels and process cooling units for computer room and equipment cooling.

The 60-Hz UPS consists of three 500-kVA parallel-redundant static modules fed from the UPS input/output switchgear. The battery is rated for 15 min, 537.5 Ah. The 415-Hz UPS consists of two parallel-redundant low-acoustic-noise 75-kVA MG sets located adjacent to the computers. These sets are powered from one of the UPS 60-Hz distribution panels.

UPS 60-Hz power is delivered to the computer equipment from 14 power distribution units located on the raised floor. Each unit includes a shielded 480-208Y/120-V transformer, primary overcurrent protection, and a secondary circuit breaker panel board. Power is distributed under the floor in flexible conduit feeder assemblies.

9.8 Summary

Four examples of computer centers are given; they illustrate the use and interconnection of the components described in the preceding chapters. On the one hand, there is a similarity in the power systems for computer centers. On the other hand, each system differs from the others because of the particular physical location, inventory of com-

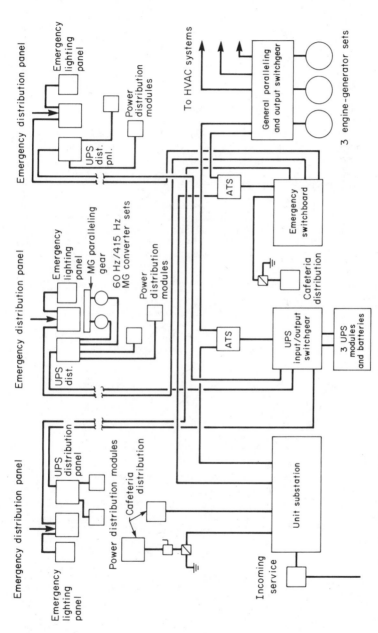

Figure 9.11 One-line diagram of electrical system for Amoco computer center [12].

143

puter equipment, and mission for the center. Unlike the conventional electrical system in the building that serves lighting, heating, A/C, and other loads, the emergency/standby systems require considerable analysis before the physical system is designed to ensure that the reliability will be achieved.

REFERENCES

1. A. Kusko, "The Quality of Electric Power," *IEEE Trans. Ind. and Genl. Appl.*, vol. IGA-3, no. 6, November/December 1967, pp. 521–524.
2. "Guideline on Electric Power for ADP Installations," *FIPS PUB 94*, U. S. Dept. of Commerce, Sept. 21, 1983.
3. "Uninterruptible Power Supply Planning Manual," *Pub. No. GA34-03160-0*, IBM, June 1986.
4. "IBM System/370, Installation Manual, Physical Planning," *Pub. No. GC22-7004-14*, IBM, June 1985.
5. UL Std. 478, 1982, "Electronic Data-Processing Units and Systems."
6. ANSI/NFPA 70-1987, "National Electrical Code®."
7. ANSI/NFPA 110-1985, "Emergency and Standby Power Systems."
8. ANSI/IEEE Std. 446-1987, "IEEE Recommended Practice for Emergency and Standby Power Systems for Industrial and Commercial Applications."
9. A. Berutti, "A Systems Approach to Computer Power," *EC&M*, January 1986, pp. 67–71.
10. R. B. Caine, "UPS Protects Fed's Computers," *Electrical Consultant*, May/June 1986, pp. 28–36.
11. R. J. Lawrie, "Designing and Installing a Power System for a Large Computer Center," *EC&M*, January 1986, pp. 76–83.
12. J. B. Esmond and P. A. Petska, "UPS and Engine-Generator Sets Protect Amoco Computer Center," *Electrical Consultant*, March/April 1985, pp. 8–59.

10

Health Care Facilities

Health care facilities include hospitals, nursing homes, and life care centers. These facilities are becoming increasingly dependent upon electrical equipment for patient life support and treatment, as well as data processing equipment for patient treatment, record keeping, and business management. A highly reliable supply of emergency/standby power is required to ensure that the lives of sick or disabled persons are protected in the event of utility power supply failure.

Numerous codes govern the essential electrical system requirements for health care facilities. The terminology, to be described in Sec. 10.2, is different from that used for computer centers and other emergency systems. Examples of essential electrical systems for health care facilities will be given in this chapter.

10.1 Standards

The standards governing health care facilities include the following:

1. ANSI/NFPA 99-1984, "Standard for Health Care Facilities," Chap. 8, "Essential Electrical Systems for Health Care Facilities" [1]

2. ANSI/NFPA 70-1987, "National Electrical Code®,"* Art. 517, "Health Care Facilities" [2]

3. ANSI/NFPA 110-1985, "Standard for Emergency and Standby Power Systems" [3]

4. ANSI/IEEE Std. 602-1986, "IEEE Recommended Practice for Electric Systems in Health Care Facilities" (White Book) [4]

---

*National Electrical Code® and NEC® are registered trademarks of the National Fire Protection Association, Inc., Quincy, Mass.

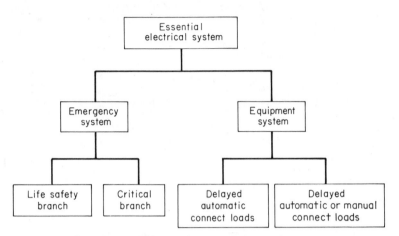

Figure 10.1   Organization of hospital essential electrical system.

10.2 Definitions

The definitions of the various parts of the essential electrical systems for hospitals and nursing homes are given in the National Electrical Code® [2]. The basic definition, from Art. 517-2, is as follows:*

> Essential Electrical System. A system comprised of alternate sources of power and all connected distribution systems and ancillary equipment, designed to assure continuity of electrical power to designated areas and functions of a *health care facility* during disruption of normal power sources, and also designed to minimize disruption within the internal wiring system. [Emphasis added.]

The Code further differentiates between hospitals and nursing homes. The diagrams of the essential electrical systems and the definitions for the various portions are given as follows.

Hospitals

A block diagram of the essential electrical system for a hospital is shown in Fig. 10.1. The definitions follow:

Essential electrical systems, general
[Art. 517-60(a)]

> (1) Essential electrical systems for *hospitals* shall be comprised of two separate systems capable of supplying a limited amount of lighting and

---

*Reprinted with permission from NFPA 70-1987, National Electrical Code®, Copyright©1986, National Fire Protection Association, Quincy, MA 02269. This reprinted material is not the complete and official position of the NFPA on the referenced subject which is represented only by the standard in its entirety.

power service which is considered essential for life safety and effective hospital operation during the time the normal electrical service is interrupted for any reason. These two systems shall be the *emergency system and the equipment system.* [Emphasis added.]

(2) The emergency system shall be limited to circuits essential to life safety and critical patient care. These are designated the *life safety branch and the critical branch.* [Emphasis added. Article 517-61 requires that all functions be restored within 10 s after interruption of the normal source.]

(3) The equipment system shall supply major electrical equipment necessary for patient care and basic hospital operation.

Life safety branch (Art. 517-62)

The life safety branch of the emergency system shall supply power for the following lighting, receptacles and equipment:

(a) Illumination of Means of Egress...
(b) Exit Signs...
(c) Alarm and Alerting Systems...
(d) Communication Systems...
(e) Generator Set Location...

Critical branch (Art. 517-63)

(a) Task Illumination and Selected Receptacles. The critical branch of the emergency system shall supply power for task illumination and selected receptacles serving the following areas and functions related to patient care.

(1) Anesthetizing locations...
(2) The isolated power systems...
(3) Patient care areas...
(4) Additional specialized patient care...
(5) Nurse call systems
(6) Blood, bone and tissue banks
(7) Telephone equipment room...
(8) Task illumination, receptacles, and special power circuits...
(9) Additional task illumination, receptacles and power circuits...

Nursing homes

A block diagram of the essential electrical system for a nursing home is shown in Fig. 10.2. The definitions follow:

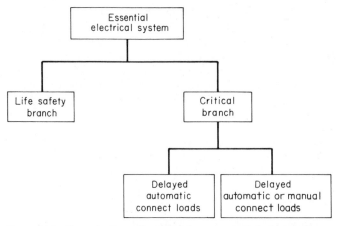

Figure 10.2   Organization of nursing home essential electrical system.

Essential electrical systems, general
[Art. 517-44(a)]

Essential electrical systems for nursing homes and residential custodial care facilities shall be comprised of two separate branches capable of supplying a limited amount of lighting and power service which is considered essential for the protection of life safety and effective operation of the institution during the time normal electrical service is interrupted for any reason. These two separate branches shall be the *life safety branch and the critical branch....* The essential electrical system shall...be so installed and connected to the alternate source of power that all functions specified herein shall be automatically restored to operation within 10 seconds after the interruption of the normal source. [Emphasis added.]

Article 517-45 requires that all functions for the life safety branch shall be automatically restored within 10 s. Article 517-46 requires that all functions for the critical branch shall be automatically restored at appropriate time-lag intervals.

10.3   Example: Typical Hospital Wiring Arrangement [1]

Figure 10.3 is a simplified schematic diagram of a hospital electrical system. The utility power is supplied by two feeders, a preferred and an alternate; switching is done on the secondary side of the transformers. One or more engine-generator sets serve as the alternate source of power. The essential equipment load is divided by priorities, by the allowable delay in restoring each priority of load, and the allowable shedding of each priority in the event an engine-generator set does not start.

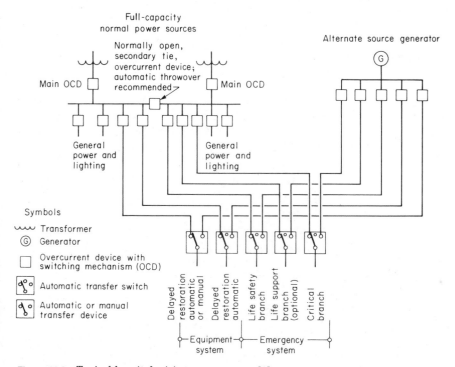

Figure 10.3 Typical hospital wiring arrangement [1].

10.4 Example: Nursing Home Wiring Arrangement [5]

Figure 10.4 shows the essential electrical system of a combined 120-bed nursing home and assisted-life facility, 480 "first class" independent condominium units, and other shops, offices, and recreational facilities. The alternate source for the system is a 175-kW diesel engine-generator set. The essential system consists of two subsystems: (1) life safety system and (2) critical system. The normal system for the facility is supplied from a 4000-A main distribution panel.

10.5 Example: Hospital System [6]

The simplified one-line essential system diagram for a 501-bed, acute-care, multispecialty hospital is shown in Fig. 10.5. The system is supplied by a normal 480-V bus fed from the utility 4.16-kV system and from an alternate 480-V bus fed from the utility 12-kV system. In addition, two 750-kW diesel engine-generators provide a 480-V emergency source. Two dual transfer switches supply the critical branch and the life safety branch. The transfer switches can supply their

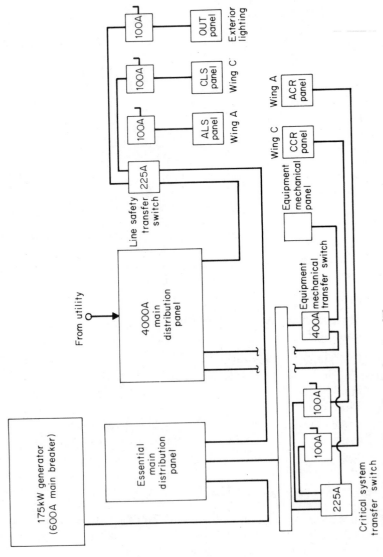

Figure 10.4 Nursing home electrical system [5].

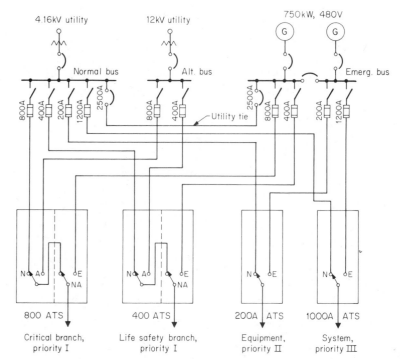

Figure 10.5   One-line diagram of hospital emergency system.

loads from the normal bus or the alternate bus or from the emergency bus. Two additional transfer switches supply the equipment system load from the normal bus or the emergency bus.

The two diesel engine-generators also provide peak-shaving power during nonemergency conditions. A utility tie with 2500-A circuit breakers is provided between the emergency bus and the normal bus.

10.6 Summary

The emergency/standby electric power requirements for health care centers are defined by the NEC® and NFPA codes. A UPS may be superimposed on the code system to supply data processing equipment, as well as critical medical instrumentation. The engine-generator sets of the code systems can back up the UPS. In a large hospital, the resultant overall system will be highly complex.

REFERENCES

1. ANSI/NFPA 99-1984, "Standard for Health Care Facilities."
2. ANSI/NFPA 70-1987, "National Electrical Code®."

3. ANSI/NFPA 110-1985, "Standard for Emergency and Standby Power Systems."
4. ANSI/IEEE Std. 602-1986, "IEEE Recommended Practice for Electric Systems in Health Care Facilities."
5. M. O'Neal, "Systems Help Life-Care Center Achieve New Levels of Luxury," *CEE,* October 1986, p. 22.
6. L. Barrett, A. Kats and N. Klaucens, "Engine Generators Shave Peak Loads and Costs at Hospital," *Electrical Consultant,* January/February 1986, pp. 26–34.

Office Buildings

Modern office buildings and complexes are cities unto themselves. They include shops, offices, living areas, garages, connections to public transportation, and other accommodations for 24-h occupancy. Emergency/standby electric power facilities are governed by state, city, and NFPA codes. In addition, power for such special facilities as computers, telecommunications, and laboratories may be provided by emergency/standby systems of motor-generator sets and UPSs. Two examples of the designs of such systems in large office buildings will be given in this chapter.

11.1 Building Loads

Loads that must be supplied by emergency/standby systems include the following:

1. *Lighting.* Emergency lighting systems are governed by ANSI/ NFPA 70-1987 [1], ANSI/NFPA 101-1985 "Life Safety Code" [2], and local codes. The types of systems include:

   - Unit lighting equipment using self-contained 6- or 12-V battery units for hallways, stairwells, and equipment rooms.

   - Central battery lighting systems utilizing a central 32- to 115-V battery, charger, and fused distribution panels. Such systems can also supply alarm and protection circuits, facilities testing and maintenance equipment, and monitoring equipment.

   - Multiple-source systems using contactors to energize emergency lights when the normal source fails. Power is taken from batteries or from the emergency ac power system for the building.

2. *Elevators.*   Emergency power is required to:

- Clear stalled elevators
- Provide adequate elevator service to clear a building during a complete utility power failure.
- Provide minimum service until utility power is restored.

3. *HVAC.*   Engine-generators provide adequate emergency power for the following types of loads. HVAC controls and control computers may require a UPS in both normal and emergency operation to provide clean continuous power.

- *Building heating and cooling for personnel.*   Restoration of power in up to 30 min is adequate, but shorter times may be required to prevent freezing or to ensure continuation of normal activities in the building.
- *Ventilation for personnel.*   Particularly in buildings with sealed windows, ventilation should be restored within 1 min. Cooling services for data processing facilities require more rapid restoration than for offices.

4. *Fire protection.*   Fire protection systems must be designed to detect and alarm potential fires and then ensure operation of fire-fighting apparatus if a fire does occur. All of these requirements are governed by state, city, and NFPA codes. Factors include the following:

- *Alarms, annunciator.*   Must be restored within 1 s; circuits are usually battery-operated locally and centrally.
- *Fire pumps.*   Must be restored within 10 s. Electric power circuits to fire pumps must be routed and protected from fire damage.
- *Auxiliary lighting.*   Must be restored within 10 s to provide light for operating equipment and guiding fire-fighting personnel.

5. *Data processing and communications.*   Data processing equipment requires UPS support backed up by engine-generators for long-time power outages. Modern computer-type telephone equipment, unless provided with internal batteries to preserve memories and other functions, requires UPS support. Conventional communications equipment, such as teletypes, internal telephones, closed-circuit TV, radio, intercoms, and paging, require power within 10 s after power failure.

6. *Mechanical utilities.* Pumps for cooling, drinking, and sanitary water, ventilation fans and blowers, and boiler auxiliaries require power in 10 s or more. Such motor-driven equipment must be restarted sequentially in order of need and function to prevent overloading engine-generators. Controls and critical instrumentation must be supported by UPS.

11.2   Example: PPG Headquarters [3]

PPG Industries in downtown Pittsburgh includes a 40-story tower, a 14-story annex, and four 6-story buildings. From an electrical point of view, the tower is not a 40-floor structure; it is two 20-story buildings stacked one on top of the other. The building and its mechanical systems are supplied with power from two utility-owned transformer stations. One at the top using four 1000-kVA transformers supplies the upper 20 stories; another at the bottom using four 200-kVA transformers provides power for the lower 20 floors. High-voltage 23-kV utility cables provide power to each of these transformer stations.

Figure 11.1 is a simplified one-line diagram of the emergency power system including the 60- and 415-Hz UPSs. The normal power source is a 2000-A 480Y/277-V distribution panel fed from one of the tower transformer stations.

The 40-floor structure is served by two standby 900-kW engine-generator sets. The first of these is dedicated to critical systems—life safety, fire pump, exit lighting, and elevator; the second supports the UPS and air-conditioning systems for the computer. The generators are interconnected by a tie breaker so that the second will take over critical functions if the first fails.

The emergency generator room also houses the emergency distribution switchgear, starting batteries, and a 500-gal fuel day tank. In the basement of another building, an 8000-gal main oil tank can provide fuel for 2 days to both generators operating at full load. From the generator room, four sets of cables, each set consisting of four 500-kcmil conductors in a 3½-in conduit, are routed to one of the fifth-floor tower core electric closets, where they connect to a 2000-A emergency riser bus duct and emergency feeders. On each office floor, the emergency bus duct riser and emergency panel are in an electrical closet different from the closet for the normal power supply bus ducts and major panels.

Located within the security-protected area of the PPG data center on the fifth floor is the air-conditioned UPS room containing two 500-kVA (0.9 PF) 60-Hz UPS modules and four 75-kVA (0.9 PF) 415-Hz UPS modules, as well as associated transfer switches and input-output switchboards. Fifteen-minute capacity lead-calcium batteries

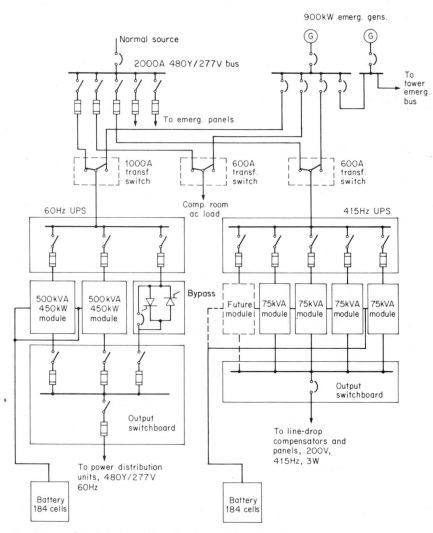

Figure 11.1 Simplified one-line diagram of PPG headquarters emergency power system [3].

are located two floors directly above in a UPS battery room. The air in this room is exhausted directly to the outside and is maintained at 77°F for long battery life. There are two sets of eight battery racks, two tiers high, for both the 60- and 415-Hz systems.

The 60-Hz 1000-kVA UPS system has feeder and fuse protection rated at 1000 A, which equates to 830-kVA continuous operating capacity. This is well above forecasted 60-Hz computer loads of approximately 750 kVA. Present 60-Hz demand loads require the operation of only one module. In fact, if both modules are operated to serve the present eight power distribution units (PDUs), the additional power loss of the second module is calculated at 100 kW and the energy costs would be $79,000 more per year.

From the 60-Hz critical load distribution center, 480-V three-phase conduit and cable feeders are routed within the raised floor to PDUs located among the computer equipment. Each PDU sitting on the raised floor contains a main primary three-pole circuit breaker, a double-shielded three-phase 480-208Y/120-V transformer, a main secondary circuit breaker, branch circuit breakers with power monitor, and surge suppressors. A shunt trip device on the primary breaker shuts down the electronic equipment when the exit door emergency shutdown buttons are pressed.

Power monitors indicate input-output voltages and amperes and kVA loading on each unit. High transformer temperature is indicated by an alarm. From each PDU a flexible conduit feeder extends in the raised floor to each piece of computer equipment served by the PDU and terminates in a grounded receptacle into which the machine is plugged. Separate ground wires are routed from the PDU ground bus to the receptacle ground screw. All conductors are coded (black, red, blue, white, or green) to ensure proper receptacle phase orientation. If the computer equipment is relocated within the confines of the data center, the associated PDU also will be relocated.

The 415-Hz 300-kVA UPS has fuses and feeders sized at 760 A, or 263 kVA, and permits additional 415-Hz modules to be installed if needed. The existing system accommodates present and projected 415-Hz computer loads which total about 200 kVA. Present loading on this 415-Hz UPS system warrants the operation of only three modules. The fourth module is a redundant spare; an annual energy savings of $15,000 is realized by not operating the fourth module. Line drop compensators are provided in the 415-Hz output cabinet to compensate for inductive voltage drop in feeder runs to the IBM 3037 power control system cabinets.

11.3   Example: Dow Jones Offices [4]

The electric power system for the new editorial offices of *The Wall Street Journal* and other Dow Jones financial services in the World Financial Center, New York, included the requirements that certain lighting, HVAC, and computer communications equipment be operational regardless of any problems that might occur with the serving utility, even an extensive power outage. Since the facility was to contain a communications center that constantly receives business information from around the world and daily transmits newspaper copy via satellite to regional printing plants throughout the United States for daily printing of the *Journal,* a system for emergency operations had to be designed into the power distribution system. Five discrete power distribution systems were installed within the nine floors occupied by Dow Jones. These systems are shown in Fig. 11.2.

1. *Base building power system.* Provides power to the air-conditioning units and to several standby distribution panels (SDP) on each floor. This base power system is served from the utility network that delivers power throughout the Financial Center complex for such equipment as elevators, the general building air-conditioning equipment, and tenant power. Power is received at 13.2 kV,

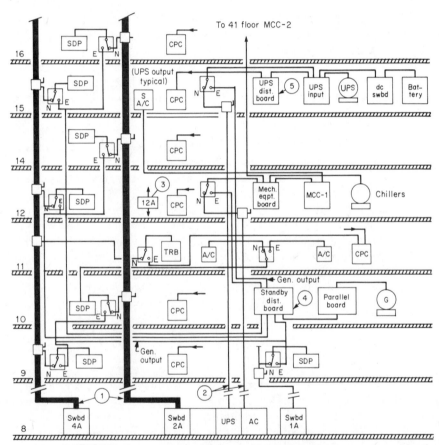

Figure 11.2 One-line diagram of the five discrete power distribution systems for the nine floors occupied by Dow Jones & Co. They are (1) base building power system, (2) dedicated Dow Jones power system, (3) base building emergency system, (4) dedicated standby diesel generator system, and (5) uninterruptible power supply (UPS) system. SDP = standby distribution panel; CPC = computer power panel [4].

transformed, and carried upward in bus ducts from switchboards 2A and 4A. It serves loads at voltages of 480Y/277 V and 208Y/120 V. This power source is subject to any and all power failures experienced by the utility.

2. *Dedicated Dow Jones power system.* Consists of two main 480-V cable risers deriving their source from the same point as the base building power system. One serves as the normal utility input to the twelfth-floor automatic transfer switch for the supplementary refrigeration equipment room; the second is the normal utility input to the fifteenth-floor automatic transfer switch for the uninterruptible power supply (UPS).

3. *Base building emergency system.* Serves only exit signs and minimal egress lighting on the Dow Jones floors. These are served from two panels in the twelfth-floor electrical closet A. Other elements served by this system are the stair lights, selective elevator control, and building alarm systems.

4. *Dedicated Dow Jones standby diesel generator system.* Routed through and serves virtually all Dow Jones office spaces and thus permits certain data processing equipment, lighting, and air-conditioning equipment to be operational even during a prolonged loss of utility power. This system utilizes either the base building power system or the Dow Jones dedicated power system as its normal power input. See Fig. 11.3.

In the event that all or part of these sources fails, the respective transfer switches signal the three 1200-kW generators to start. The load is transferred by each affected transfer switch when the generating power becomes available. Panels connected to this system are encoded as follows:

- SDP designates the *standby distribution panel* and is preceded by the floor number. This panel serves as the distribution point for all other standby distribution panels on the floor.

- SHV, the *standby 480/277-V lighting panel* taken off the SDP, generally serves the 277-V lighting critical to office operations.

- SLV, the *standby 208/120-V panel,* is taken off the SDP and generally serves furniture-mounted indirect lighting powered via the raised floor system and such miscellaneous equipment as copiers that are critical to operations but are not subject to damage due to power fluctuations or momentary outages.

- SAC is the *standby air-conditioning panel,* also taken off the SDP. It serves floor-mounted (and some ceiling-mounted) supplemen-

Figure 11.3 Control panels for three 1200-kW diesel-generator sets, Dow Jones & Co. [4].

tary air-conditioning units arranged to provide adequate environmental conditions for critical equipment or spaces. These include the packaged units on each floor and the units in the computer room and data/communications equipment room.

Packaged air-conditioning units located in closets or alcoves adjacent to the office areas permit a tolerable comfort level to be maintained for personnel in these occupied spaces if utility power is lost for an extended period. The standby air-conditioning power system also provides the emergency input to the transfer switches for the supplementary refrigeration plant and the UPS.

5. *Uninterruptible power supply (UPS) system.* Receives its input from a transfer switch and uses motor-generator sets. The system consists of three paralleled 60/60-Hz 360-kVA motor-generator units located on the fifteenth floor. The UPS units, in turn, are supported by a battery system that can provide power for 15 min under full load to the rotating equipment in the event of utility failure. Output of the UPS system is directed to a switchboard containing 11 feeder circuit breakers, each of which serves an individual computer power center (CPC) that delivers power to localized loads. A 480-208Y/120-V shielded isolation transformer in the CPC supplies

equipment utilization voltage, through circuit breakers and branch circuits, to individual computers or peripherals. See Figs. 11.4 and 11.5.

A contingency capability is achieved by three basic methods:

1. *Redundant equipment.* All components of a system operate simultaneously but are oversized so that, should a component fail, the remaining equipment can support the load. The utility service to the Financial Center complex operates in this manner. For example, a single utility feeder and transformer can fail simultaneously, and the network is still able to support design load. All electrical equipment is sized to accommodate a 24 percent additional load above design calculations for up to 2 h. This is in addition to a planned spare capacity of at least 20 percent.

2. *Alternate source.* During a failure condition, a replacement for the failed component can be installed. The most common way to implement this method is through the use of automatic transfer switches.

3. *Substitution of a failed component.* This procedure provides for an alternate source but only after manual inspection and review determine the appropriateness of the transfer. The equipment used for this method includes manual transfer switches, maintenance bypass breakers, and isolation bypass switches. The plan includes, for example, having an extra CPC in storage should the transformer or some other component in a CPC fail. Thus, either a part or a complete computer power center can be substituted.

Figure 11.5   UPS distribution panel, Dow Jones & Co. [4].

11.4 Summary

The emergency/standby electric power system for an office building must meet the code requirements for emergency lighting, emergency elevator operation, emergency ventilation, as well as UPS power for specialized computer, communication, and instrumentation equipment. The installation of battery rooms, engine-generator sets, and fuel tanks must also comply with national and local codes.

REFERENCES

1. ANSI/NFPA 70-1987, "National Electrical Code®."*
2. ANSI/NFPA 101-1985, Life Safety Code.
3. M. R. Marxreiter and E. D. Piazza, "PPG Headquarters Assured Dependable Power," *Electrical Consultant,* March/April 1986, pp. 8–16.
4. J. R. Knisley, "Dependable Power Serves Dow Jones Offices," *EC&M,* January 1986, pp. 53–61.

---

*National Electrical Code® and NEC® are registered trademarks of the National Fire Protection Association, Inc., Quincy, Mass.

12

Remote Sites

Remote sites are not expected to have normal utility electric power service. Equipment located at typical sites include microwave repeaters, fiber-optic cable repeaters, satellite receiver stations for telecommunications, and instrumentation packages. The requirements for the electric power supply for an operating system at a remote site include the following:

1. Power must be available on demand 24 h/day, 365 days/year.
2. Continuous automatic operation of the system must be assured.
3. Power system reliability must equal or exceed that of the system equipment.
4. Maintenance needs should be infrequent and straightforward.
5. Fuel consumption should be low or zero.
6. Pollution should be minimized.
7. Equipment and running costs should be as low as practically possible.

Examples of three power-supply systems at remote sites will be shown.

12.1 Types of Electric Power Supplies [1]

The types of electric power supplies for remote sites where no utility power is available are shown in Table 12.1. The delivered power ranges from 10 W to 200 MW.

TABLE 12.1   Characteristics of Remote Site Electric Power Supplies [1] (©1985, IEEE)

Generator type	Continu-ous load range	Maintenance and reliability	Fuel use	Average life-time, years	Installation
Photovoltaic power system	10 W to 6 MW	Annual checks; high reliability; > 100,000 h MTBF	Nil	15 to 20 (esti-mated)	Simple
Wind generator	50 W to 1 MW	Frequent checks; poor reliability	Nil	5 to 10	Difficult
Gasoline gener-ator	1.0 to 2.5 kW	Frequent checks; low-duty cycle	High	2 to 3	Simple
Diesel gener-ator	3.5 kW to 200 MW	Every 1–3 months; regular overhauls; 3000–8000 h MTBF	High	3 to 5	Difficult
Closed-cycle vapor turbine	0.4 kW to 3.0 kW	Annual checks; > 20,000 h MTBF	Med-ium	10 to 15	Difficult
Thermoelectric generator	10 to 120 W	Annual checks; > 30,000 h MTBF	High	15 to 20	Difficult

Engine-generators

For dc loads greater than 200 W continuous, diesel generators have generally been installed in telecommunications systems. Above 1000 W continuous loads, two (or even three) diesel generator sets are run alternately and supply the load through duplicate sets of rectifiers and a standby battery.

The smallest low-speed diesel generators available today are rated at approximately 3 kW; thus they are usually oversized for a telecommunications application. This light loading may lead to increased maintenance and shortened life because of carbonization of exhaust valves and injectors and cylinder bore glazing. With the continuing trend of reduced energy consumption by telecommunications equipment, below 200 W dc, even if diesel generators are run infrequently at full load (to charge batteries), a 3-kW unit is not economic in terms of life cycle cost. Because of the light loads involved, high maintenance and operating costs become the rule rather than the exception.

Batteries

Photovoltaic power systems depend upon batteries to provide power continuously at night and during periods of low illumination of the solar arrays. Suitable requirements for the batteries are the following:

1. Good tolerance of deep and shallow discharge cycles; 1200 cycles at 80 percent depth of discharge; 3000 cycles at 50 percent depth of discharge.

2. Low self-discharge, maximum 2 percent/month

3. High charging efficiency, greater than 97 percent up to 50 percent state of charge

4. Low electrolyte losses and large electrolyte capacity to minimize maintenance

5. Rugged mechanical construction

The requirements can be approached with lead-calcium and pure lead batteries (cycle life limited), automotive batteries (also cycle life limited), and traction batteries (high water loss and high self-discharge).

Regulators

In photovoltaic power systems the battery stack, when required, receives energy sufficient to meet future load demand but limited to prevent overcharging, loss of electrolyte through gassing, and subsequent damage. Thus, a regulator is required to channel power from the photovoltaic array to charge the battery.

Early solar power regulators were of the linear shunt design; charging current was limited by a voltage sensor which diverted excess current from the battery to a shunt resistor, where the power was rejected as heat. This had the disadvantage that the heat sink had to be very large, especially in hot climates, which limited the system to low-power applications—maximum 100 W peak. To avoid heat dissipation, switching-type regulators were introduced. Some utilize shorting transistors placed across the output of each array section and controlled by battery voltage-sensing circuitry.

Three additional features are essential for reliable operation in remote communication power systems:

1. Temperature compensation is required to reduce charging voltage in proportion to temperature; otherwise, undesirable gassing would occur and eventually lead to battery damage.

2. Dual voltage setting first takes battery voltage higher than normal to equalize cell voltage after periods of no charge, usually at night. This mixes the electrolyte to avoid any local concentration which may cause plate corrosion. Charge voltage then automatically resets to a low float level to maintain charge during the normal solar day.

3. Alarms and output controls are included to indicate low or high battery voltage. Each alarm can be extended into a telemetry system. An automatic load-disconnect device operates after the low-voltage alarm to protect the battery or load from dangerously low voltage.

12.2   Example: Photovoltaic Power System for Telecommunications [1]

A partially redundant high-reliability system is shown in Fig. 12.1. The single solar array is split into four subarrays to support the design load via the battery bank. Solar array power is controlled by two parallel electronic charge regulators (ECU) with their respective sets of high- and low-voltage alarms, system status metering, and automatic low-voltage load-disconnect devices. The charge regulators supply array power to two battery banks, each of which has the capacity to support the load for half of the required storage period. A wall-mounted socket is provided to allow connection of an external battery charger during maintenance.

Should one half regulator fail, the remaining regulator is sufficiently rated to supply the total array current. Both battery banks, A and B, are connected in parallel to the remaining healthy regulator before a blocking diode; thus both are kept in a high state of charge. In this way the system is maintained in operation without interruption. The telecommunications system will continue to operate, and repair can be relegated to a "nonurgent" status, the failure being signaled by appropriate alarms connected to a supervisory system if required.

As shown in Fig. 12.1, a total of four switching stages is used, thereby providing closer control over the array output. Thus when

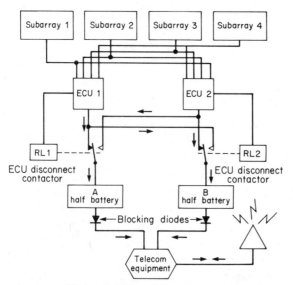

Figure 12.1   High-reliability photovoltaic power systems [1]. (©*1985-IEEE*).

battery charge is low, all four sections are connected (producing up to 80 A by relays, more with contactors); and as the battery becomes charged, sections are cut out to finally reach open circuit. The use of relay switching elements is preferred to solid-state units for several reasons:

- Relays switch approximately once every 15 min (when regulating only) and cause no radio frequency interference, as opposed to the high frequencies employed in some solid-state regulators.

- Series power loss is low, so charging efficiency is high.

- Relay switches are reliable under transient loads.

- A relay sequential switch readily approximates the linear battery charging characteristic.

- Maintenance is simplified, since faults are more easily traced.

12.3 Example: Repeater for Optical Fiber Cable [2]

Repeaters are required every 20 km for a 500-km 550-kV transmission line in Japan, which employs composite overhead ground wires with optical fiber cable. Each repeater requires 20-W power at 24 V dc. Four methods are shown in Table 12.2: (1) solar power, (2) wind power, (3) fuel cell, and (4) engine-generator. Based on cost, convenience and reliability, the solar power system has been selected.

The solar power system must deliver power continuously for up to 20 days without sun. The specifications for the solar cell modules and batteries are given in Table 12.3. The solar cell array consists of eight modules, each with a peak power of 35 W and an area of 0.37 m^2. The battery consists of eleven 800-Ah lead-acid cells; the cell voltage declines from 2.35 V/cell at full charge to 2.0 V/cell at discharge of about 420 Ah.

The power supply circuit is shown in Fig. 12.2. The battery charging and bus voltage is regulated by a shunt-transistor regulator. The complete equipment with the solar cell array mounted on top is shown in Fig. 12.3.

12.4 Example: Small Earth Station for Satellite [3]

Figure 12.4 shows a diagram of a satellite system in West Germany that will ultimately employ 100 small earth stations and interconnect

TABLE 12.2 Comparison of Methods for Supplying 20 W [2]

Power supply method	Basic circuit	Outline and cost of equipment	Advantages	Disadvantages
Solar power	Voltage regulator, Reg., Solar cell, Batteries	• Necessary area for solar cell: 3 to 4 m^2 (8 to 11 modules) • Battery capacity required: 24 V 400 Ah to 1000 Ah (for 20 days without sunlight) • Cost rate: 1	• Few physical limitations on location • No moving parts makes maintenance easy • Easy to operate • Low cost in small-capacity unit	• Needs batteries • Solar cell field requires a fairly large area • High cost in large-capacity units
Wind power	Windmill generator, Reg, Field regulator, Batteries	• Wind area received by windmill (much larger than area where windmill is set up): 0.5 to 3 m^2 • Battery capacity required: 400 Ah to 1500 Ah (for 5 days without wind) • Cost rate: 1.1 to 6.5	• No fuel needed • Generates electricity even at night	• Controlled by weather • Safety measures for strong winds required • Rotating parts make maintenance necessary • Battery needed • Power cannot be generated at wind speeds of less than 3 m/s

TABLE 12.2 Comparison of Methods for Supplying 20 W [2]   (*Continued*)

Power supply method	Basic circuit	Outline and cost of equipment	Advantages	Disadvantages
Fuel cell generator (alkaline electrolysis fluid)	Fuel electrode H_2 O_2 (Air) Hydrogen cylinder	• 15 days operation with 7-m^3 cylinder • Spare cylinder and automatic switching valve necessary • Cost rate: 2 to 2.5	• Routine maintenance easy because there are no moving parts • No poisonous fumes emitted • No batteries needed • Setup area small	• Difficult to operate • A supply of pure hydrogen is required • There are difficulties in controlling electrolytic fluid density • Fuel and electrolysis systems need to be monitored • Fuel cell life of 10,000 h
Engine generator	Engine generator G — Engine Fuel tank	Note: There are no small-scale portable systems, and none are reliable in continuous long-term operation • Cost rate: 0.8 to 2.5	• Cheap and compact	• Periodic fuel replenishment necessary • Periodic maintenance of moving parts necessary • Gas emitted from internal-combustion engine • Only generators of 500 W and more on market

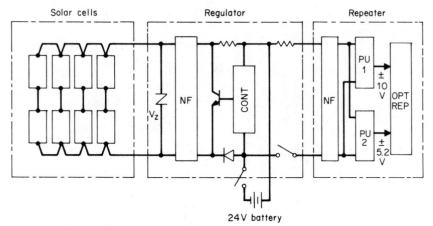

Figure 12.2   Circuit diagram of solar cell power supply [2].

NF	Noise filter
REP	Repeater
CONT	Charge and discharge control circuit
PU1, PU2	Power units
V_Z	Varistor
OPT REP	Optical repeater

TABLE 12.3   Specification for the Solar Cell Modules and Batteries [2] (©1985, IEEE)

Item	Specification
Solar cell module	ELR604-160 Maximum output: 35 W Efficiency: 9.5% Area: 0.37 m² (Fuji Electric Co., Ltd.)
Number of modules	8
Supplied voltage	24 V dc ± 10%
Number of battery cells in series	11
Battery voltage fluctuation	2.35 V × 11 = 25.85 V* 2.0 V × 11 = 22 V†
Battery cell capacity	800 Ah, PS-800TL (lead-acid battery) (Furukawa Battery Co., Ltd.)

*Voltage to prevent overcharging.
†Voltage when storage cell expires.

Figure 12.3 Building housing the solar cell power supply equipment [2]. (*Courtesy, FUJITSU LTD.*)

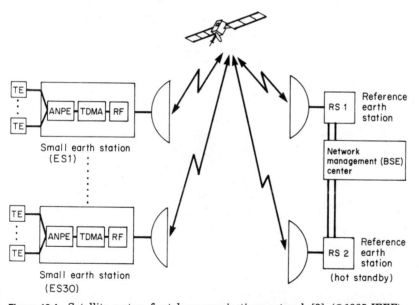

Figure 12.4 Satellite system for telecommunications network [3]. (©*1985-IEEE*).

ANPE	Digital exchange
RF	Satellite radio equipment
TE	Terminal equipment
TDMA	Time-division multiple-access equipment

with customers' terminal equipment by land lines. When the small earth station is installed at a remote site in a container, 380Y/220-V three-phase utility power is supplied. When the equipment is installed in a building of the telecommunications systems, 220 V ac is required for the TDMA equipment and 60 V dc for the ANPE. Only a 60-V dc UPS (batteries) is available. A UPS for the small earth stations requires additional investment. The purpose of this analysis is to compare the availability of a channel between two small earth stations with and without UPS.

The power supplies of the small earth station equipment can tolerate a utility power interruption of up to 20 ms. For longer interruption times, the total downtime of one station is given by:

$$B_{ES} = B_{PS} + I_{RF} + I_{TDMA} + I_{ANPE}$$

where B_{ES} = downtime of the small earth station
B_{PS} = downtime of the power supply unit
I_{RF} = initialization time of RF
I_{TDMA} = initialization time of TDMA
I_{ANPE} = initialization time of ANPE

The necessary initialization times are summarized below:

Technical equipment	Utility failure duration	Initialization time
RF	> 20 ms	1–5 min
TDMA	> 20 ms	2 min
ANPE	> 20 ms	3 min

The utility system was monitored at 60 locations. The interruptions longer than 20 ms at each small earth station site were predicted as 2.8/month. Three cases were studied, with the following results:

1. RF, TDMA, and ANPE without UPS (i.e., all components are in containers).

- Average downtime for one small earth station = 0.47 h/month.
- Reduction in availability of two stations = 0.13 percent.

2. RF and TDMA without UPS; ANPE is installed in a telephone exchange and supplied from the existing UPS with 60 V dc (standard configuration).

- Average downtime for one small earth station = 0.34 h/month.
- Reduction in availability of two stations = 0.09 percent.

3. RF without UPS; ANPE is supplied from existing 60 V dc UPS; TDMA is supplied with 220 V ac from existing UPS (exceptional case).

- Average downtime for one small earth station = 0.25 h/month.
- Reduction in availability of two stations = 0.07 percent.

The reduction in availability because of utility power failure is 0.13 percent for the remote site and 0.09 percent for the telephone exchange site. The addition of a UPS for the TDMA only changes the reduction to 0.07 percent. The decision was made to operate the small earth stations without a UPS, but the utility power reliability at proposed sites will be studied to ensure that sites with above average numbers and durations of failures are not selected.

12.5 Summary

Remote sites for repeater stations, instrumentation, and satellite receivers are expected to have no utility electric power service. They rely on photovoltaic generators, wind generators, thermoelectric generators, and small engine-generator sets. Batteries, inverters, and regulators ensure continuous power to the load. Economic studies are usually made to select the best electrical systems for a particular application.

REFERENCES

1. I. F. Garner, "Photovoltaic Power System Design for Telecommunications," *Intelec '85*, pp. 461–469.
2. M. Ohara, K. Hanyuda, T. Suzuki, and K. Watanabe, "Solar Cell Power Supply System for Composite Overhead Ground Wire Telecommunications Systems with Optical Fibers," *Intelec '85*, pp. 547–553.
3. J. Strasser, "German Communications Satellite System, Examination of the Necessity for an Uninterruptible Power Supply for Small Earth Stations," *Intelec '85*, pp. 69–72.

Procedures

Load Classification

The purpose of an emergency/standby system is to provide an alternate source to the utility power line. Loads range in sensitivity to loss of utility power from as little as one-half cycle for computers to minutes or more for heating and air-conditioning systems. Each installation comprises a range of load equipment that must be classified by sensitivity before a system is designed to supply the load in the face of utility power failure.

The three common alternate sources that can supply power to the loads include the battery directly, the UPS from its dedicated battery, and the engine-generator set. The battery can supply power with no interruption, or after about a 0.5-s transfer switch operation. The UPS can supply power with no interruption when the utility source fails. However, when a bad module is switched out of a redundant UPS, or the UPS switches to a bypass line, the power to the load can be interrupted for up to one-quarter cycle (4 ms). The engine-generator set can start, reach speed, and be able to pick up load in about 10 s. Hence, in categorizing loads, the choice of supply lies between battery, UPS, engine-generator set, or nothing, until utility power returns.

The requirements for emergency power for such facilities as hospitals, schools, public buildings, hotels, and theaters are specified by state and city building codes, the National Electrical Code®,* and NFPA Codes.

13.1 Utility Power

The path by which utility power reaches a critical load in a facility is shown in simplified form in Fig. 13.1. The pertinent electrical system

---

*National Electrical Code® and NEC® are registered trademarks of the National Fire Protection Association, Inc., Quincy, Mass.

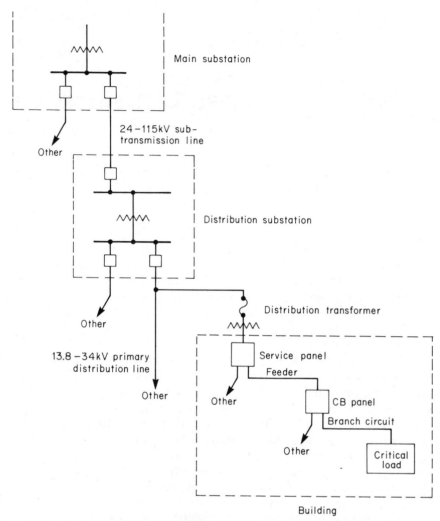

Main substation

24 – 115 kV sub-
transmission line

Other

Distribution substation

Other

Distribution transformer

13.8 – 34 kV primary
distribution line

Other

Service panel

Feeder

Other

CB panel

Branch circuit

Other

Critical
load

Building

Figure 13.1   One-line diagram of electrical system to supply a critical load.

starts in the high-voltage network of the utility, which consists of transmission lines, power plants, substations, and interconnections to other utilities. From the substation bus, several subtransmission lines extend out to supply distribution substations in each load area of the utility system. The subtransmission lines are overhead lines or underground cables ranging in voltage from 24 to 115 kV. From each distribution substation, primary distribution lines extend along the streets to supply the area's loads. These are typically three-phase, 13.8- to 34.5-kV lines which also extend as one-phase lateral lines to side streets. The commercial, industrial, and residential loads are supplied from distribution transformers located on pads adjacent to the buildings, in vaults, and on the poles. The power enters the building at a service panel, is further sent to a circuit breaker panel, and finally is sent to the critical load. The feeders within the building are typically 208Y/120 V or 480Y/277 V four-wire circuits.

Practically all power disturbances are voltage disturbances, i.e., deviations from the sinusoidal 60-Hz voltage wave. These disturbances occur in the utility system or in the building portion of the electrical system. Many investigations of the magnitude and duration of utility disturbances have been done; the disturbances are functions of many variables—company, site, season, hour, etc. See Chap. 17. They can be classified as follows:

1. *Microsecond disturbances,* transient under- and overvoltages caused by lightning and switching in the utility system.

2. *Millisecond disturbances,* undervoltages up to several cycles (50 ms) caused by capacitor switching and transformer energization in the utility system, circuit breaker and fuse blowing in the building, and cyclic loads such as welders.

3. *Second disturbances,* undervoltages caused by feeder switching in the utility system and motor starting in the building.

4. *Minute-to-hour disturbances,* caused by wide-area blackouts, overhead line failures from storms, autos hitting poles, and substation equipment failures in the utility system. Also caused by equipment failures in the building.

Typical statistical distributions of utility outages vs. frequency and length of outage, if available at a specific site, would assist a potential operator of critical load to answer such questions as (1) Is an emergency/standby system needed? (2) For what operating time must the batteries be sized? (3) Is an engine-generator required to back up battery-operated equipment?

13.2   Alternate Power Sources

The tools available to the electrical designer to protect critical equipment from utility and internal-building disturbances include the following:

1. *Line filters* to attenuate microsecond-type line-overvoltage transients which might damage equipment.

2. *Ferroresonant constant-voltage transformers* to provide voltage to

loads when the line voltage dips or overshoots for up to about one-half cycle of time (8 ms).

3. *UPS* to provide power for the time until the batteries discharge, typically up to 20 min. In normal operation, internal switching out of failed modules in the UPS can depress the output voltage up to one-quarter cycle (4 ms).

4. *Battery systems* to supply emergency lighting and other directly operated dc-power loads. Systems operate directly, or transfer in about 0.5 s, and continue for the duration of the battery charge.

5. *Engine-generator sets* to provide long-term standby power. Sets require about 10 s to start and pick up load. Duration depends upon fuel supply.

6. *Transfer switches* to provide power from an alternate utility feeder within about 0.5 s following the failure of the preferred feeder.

A typical electrical system for a facility such as a hospital or a data processing center will utilize practically all of the above equipment to supply its loads.

13.3   Load Categories

In the design of an emergency/standby system for a facility, the loads must be classified by line-voltage sensitivity and by function. Suitable power sources must be provided for each group of loads. Loads can be classified by sensitivity as follows:

1. *Critical,* requiring line voltage with less than one-quarter cycle (4 ms) of dip to zero.

2. *Essential,* requiring line voltage following a 10-s dip to zero.

3. *Nonessential,* requiring line voltage following a dip to zero lasting minutes to hours.

Loads can be classified as to function as follows:

1. *Equipment support,* necessary to operate an overall system, e.g., a data processing center, power plant boiler, industrial process, or air-traffic control center.

2. *People support,* necessary to maintain a specific group, e.g., the personnel necessary to operate an air-traffic control center or the patients and staff of a hospital.

3. *Building support,* necessary to keep a building functioning, e.g., lights, heat, air-conditioning, fire alarms, elevators.

Codes for emergency power by states and major cities are listed in Table 1 of ANSI/IEEE Std. 446-1987 [1].

13.4 Loads for Battery Supply

Two types of battery systems are used for emergency/standby supply. In the first, the system consists of an ac/dc rectifier/charger supplying a dc bus with a battery floating on the dc bus. All of the critical load is supplied from the dc bus. In the event of utility power failure, the battery supplies the load on the dc bus. When ac power is restored, the rectifier/charger supplies the load on the dc bus and recharges the battery. In the second system, the ac/dc rectifier/charger supplies a battery in float condition. When the utility power fails, a switch energizes load from the battery. When ac power is restored, the load is switched off and the rectifier/charger recharges the battery.

Typical loads for battery supply include the following:

1. *Power plants and substations,* including

 - Control circuits

 - Circuit breakers, contactors

 - Back-up lubrication pumps

 - Alarms

 - Telephone, radio

 - Emergency lights

2. *Telephone systems,* including

 - Central office equipment

 - Repeaters

 - Satellite stations

 - Subscribers' equipment

3. *Buildings,* including

 - Emergency lights

 - Fire alarms

 - Telephone, radio

 - Security

13.5   Loads for UPSs

In assigning loads to categories and to methods of supply, examples of critical loads for UPS supply include the following:

1. *Data processing,* including

   ■ Disk drives

   ■ Central processors

   ■ Terminals

   ■ Communications interface

2. *Medical electronics,* including

   ■ Monitors

   ■ Instrumentation

   ■ Heart/lung machines

3. *Industrial controls,* including

   ■ Control computers

   ■ Programmable controllers

   ■ Terminals

In addition, small UPSs up to 10 kVA are used for PCs, laboratory instrumentation, photo processors, and other load equipment whose operation would be costly to interrupt.

13.6   Loads for Engine-Generator Sets

Loads which can tolerate a 10-s delay in power after utility failure include the following, as listed in Table 2 of ANSI/IEEE Std. 446-1987 [1].

1. *Lighting,* including

   ■ Security

   ■ Warning

   ■ General

2. *Transportation,* including

- Elevators

- Escalators

- Conveyors

3. *Mechanical utilities*

- Water, cooling, general use

- Water, drinking, sanitary

- Pumps, water, sanitation, production

- Fans and blowers, ventilation, heating

4. *Space conditioning*

- Heating

- Cooling

- Ventilation

- Air pollution

- Humidity

5. *Communications*

- Telephone

- Teletype, facsimile

- Radio

- Paging

- Alarms, annunciation

Example 13.1   An engineering firm will be the sole occupant of a three-story 30,000-ft^2 building. The anticipated electrical load for the building is the following:

Nature of load	Load, kW
Lighting	
Interior	750
Exterior (parking lot, security)	25
Emergency	5

Nature of load	Load, kW
HVAC	
Air-conditioning	400
Circulating fans, etc.	50
Electronic	
Computer and peripherals	45
Telephone system	5
Services	
Shop	35
Laboratory	15
Cafeteria, etc.	50
Elevator	50
Total	1430

Allocate the load into critical, essential and nonessential components, as defined in Sec. 13.3. Determine how each component shall be supplied.

solution   The load allocation is the following:

Nature of load	Load by categories, kW			
	Normal	Critical	Essential	Nonessential
Lighting				
Interior	750		150	600
Exterior	25		15	10
Emergency	5	5	(5)*	
HVAC				
Air-conditioning	400		175	225
Circulating fans	50		50	
Electronic				
Computers and	45	45	(45)*	
peripherals				
Telephone system	5	5	(5)*	
Services				
Shop	35		25	10
Laboratory	15	10	5	
Cafeteria, etc.	50		25	25
Elevator	50		50	
Totals	1430	65	550	870

*Power to UPS.

Each component of the load shall be supplied as follows:

1. Normal load. Taking into account diversity and power factor, the normal load can be supplied from the utility by a 2500-kVA transformer feeding a 480Y/277-V 3000-A switchboard.

2. Critical load. The emergency lighting load can be supplied from battery-operated units. The computer, telephone, and laboratory critical load can be supplied from UPS modules with self-contained batteries. The input lines to the UPS will be supplied from the essential buses.

3. Essential load. The essential load will be supplied by two 500-kW diesel

engine-generator sets. The load will be allocated to two 800-A essential switch-boards as priority 1 and priority 2. Each switchboard will be supplied by a transfer switch either from utility power or from the generator bus. The two generators will be synchronized to the generator bus. If one generator is out of service or fails to start, the priority 2 load will not be transferred.

4. Nonessential load. The load will be restored when utility power is restored. If diesel engine-generator sets larger than 500 kW each are installed, more nonessential load can be categorized as essential load.

13.7 Summary

An analysis of emergency conditions at a facility, similar to a fault and failure analysis, must be carried out to determine what loads must be supplied by what emergency/standby source in what time frame. The results will also determine the ratings of the UPS, engine-generator sets, and other equipment and their costs. If one critical load is not supplied properly, the whole facility may fail during an emergency.

REFERENCES

1. ANSI/IEEE Std. 446-1987, "IEEE Recommended Practice for Emergency and Standby Power Systems for Industrial and Commercial Applications."

Reliability

The purpose of emergency/standby power systems is to provide electric power when the normal utility system fails, i.e., to provide reliable service for critical load equipment. In the design of the emergency/standby system, we are concerned, first, with the reliability of the normal utility power supply and, second, with the reliability of the equipment that will provide power when the normal supply fails.

Reliability calculations provide a numerical procedure for determining the probability that a specific equipment or system will operate without failure for a given time period. The reliability can be expressed mathematically as a probability, which is a number less than 1, that the equipment will not fail in the next hour. The reliability can also be expressed, in hours, as the mean time between failures (MTBF) of a number of identical equipment units.

Once the reliability data on the components of a system are known, the reliability of the system can be calculated in the same terms. The calculation of reliability is particularly useful in comparing the performance, as well as the cost/benefits, of various designs.

14.1 Definitions

The basis for reliability analysis is the assumption that we are dealing with a large number of identical components or systems whose failures are occurring at random. This equipment undergoes a life as shown in Fig. 14.1 [1]. A large number of components or systems start life at zero on the time scale. During the period up to time T_B, called the infant mortality period, the failures are caused by weak or defective parts or manufacturing defects. When these components are weeded out, the remaining equipment settles down to a random failure rate between time T_B and T_W, which represents the useful operating life. Finally, for times greater than T_W, the failure rate increases because of old age and wearout.

This concept of failure rate certainly applies to the small components of emergency/standby systems, such as transistors and relays. It may not apply to large equipment, such as high-power UPS modules or large engine-generator sets, which are not built identically in large quantities.

Several terms used in reliability analysis will be defined as follows for operation between T_B and T_W on the curve of Fig. 14.1:

1. *Failure rate* λ (lambda), the number of failures per hour, assuming a constant random failure rate of a mature system
2. *Mean time between failures (MTBF)*, the mean time at which failures occur in a mature system, expressed in hours; also equal to 1
3. *Mean time to repair (MTTR)*, the expected time it would take to place a failed unit back in operation, expressed in hours
4. *Availability (A)*, the per unit time that a unit is operational; also equal to MTBF/(MTBR + MTTR)

14.2 Sources of Reliability Data

Utility failure rate data have been reported in the technical literature for specific conditions. Two samples are the following:

1. Table 14.1 shows power failures measured at 24 Bell data processing sites over a 270 site-month period from the paper by Goldstein and Speranza [2]. The measurements were made at a 208Y/120-V bus in each site. Such data can be used, for example, to determine the probable time intervals between power failures lasting 1 cycle (17 ms),

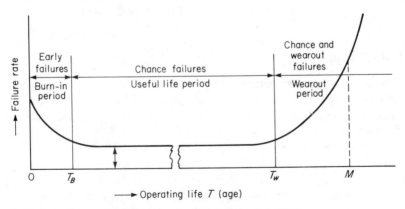

Figure 14.1   Component failure rate as a function of age [1].

TABLE 14.1   Predicted AC Line Disturbances for Power Failures Lasting Longer than the Time Indicated [2] (©1982, IEEE)

Sites having fewer than N events, %	N, the number of power failures or sags expected per year, for following minimum duration times, s,							
	0.017	0.1	0.5	1.0	10	100	1000	10,000
10	0	0	0	0	0	0	0	0
25	2	2	1	0	0	0	0	0
50	4	3	3	2	2*	1	0	0
75	7	6	6	5	4	2	2	0
90	12	10	9	8	7	6	5	2

*Of all the sites in the Bell System, 50% will experience fewer than two power failures per year, each lasting longer than 10 s.

which will cause a computer to go down, or to determine the necessity for a generator to back up a UPS with a battery capable of supplying rated UPS power for 1000 s. Such data for a specific site can be collected with a transient voltage recorder over a period of several months.

2. Table 14.2 shows electric utility failures at a group of industrial sites as obtained from replies to questionnaires solicited by an IEEE committee [3]. The failures are of the feeders to the plant. Such data are available at any site from the maintenance or utility organizations.

TABLE 14.2   Summary of "All Industry" Equipment Failure Rate and Equipment Outage Duration Data for 66 Equipment Categories Containing 8 or More Failures [3] (©1974 IEEE)

Equipment	Failure rate: failures per unit-year	Actual hours downtime per failure	
		Industry average	Median plant average
All electric utility power supplies	0.643	1.33	1.04
Single-circuit	0.537	5.66	5.10
Double- or triple-circuit, all	0.622	0.85	1.17
Automatically switched over	0.735	0.59	0.93
Manually switched over	0.458	1.87	2.00
Loss of all circuits at one time	0.119	2.00	1.58

Component and equipment failure rates are given in such sources as the following:

1. Table 14.3 from a paper by Schwarm [4]. The failure rates are

TABLE 14.3   Reliability of Various Major UPS Components [4]
(©1979, IEEE)

Component	MTBF, h	λ, failures/h
Rectifier	230,000*	4.35×10^{-6}
Inverter	50,000†	2×10^{-5}
Static bypass switch	100,000‡	1×10^{-5}
Critical circuit breakers (5)	400,000*	2.5×10^{-6}
Power and control connections (100)	690,000*	1.45×10^{-6}
Battery	280,000§	3.57×10^{-6}
Redundant or static bypass Control commonality	1,000,000¶	1×10^{-6}

*Quoted or calculated from data in Ref. 3 by using quantities typical to the system.
†Recent Arthur D. Little, Inc. field experience in which 18 large inverters performed with an MTBF of approximately 100,000 h over 134 operating years and a survey of 1000 small inverters with an MTBF of 40,000 h.
‡Estimate based upon component parts count and complexity.
§Estimate by ESB, Inc. technical staff; includes cell interconnections for a 174-cell battery. Since modern station-quality batteries usually deteriorate rather than fail catastrophically, this estimate assumes that the battery will be carefully maintained and monitored for signs of deterioration in a UPS installation.
¶Estimate based upon reduction of complexity of newer control concepts as compared to the calculated and demonstrated performance of previous common master controls with MTBFs in the range of 60,000 h.

based on Schwarm's experience with many UPS installations. Similar data are given in other technical publications.

2. Military Handbook 217D, "Reliability Prediction of Electronic Equipment" [5]. Failure rate data collected from manufacturers and from the military are tabulated in detail for electronic components. Complete equipment data are not given.

3. Failure rate data for power components such as transformers, circuit breakers, motors, and generators from the same committee that presented the data in Table 14.2 [6].

14.3   Reliability Essentials

In order to calculate the reliability of a portion or the total of an emergency/standby system, the system must be described by a reliability model. The model will consist of a group of blocks. The blocks representing the parts of a system such that any one failure will cause a system failure are connected in series. The blocks of a system such that all portions must fail to cause system failure are connected in parallel. A failure rate λ (failures/h) must be assigned to each block of the model. The system failure rate can be calculated from the model.

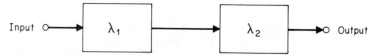

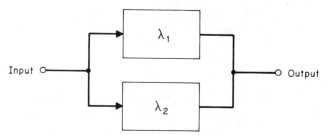

Figure 14.2 Reliability model of serial blocks.

Figure 14.3 Reliability model of parallel blocks.

The basic configurations of the blocks are shown in Fig. 14.2 for the serial arrangement and in Fig. 14.3 for the parallel arrangement. The total failure rate λ_s for the serial arrangement of Fig. 14.2 is

$$\lambda_s = \lambda_1 + \lambda_2 \tag{14.1}$$

The total failure rate λ_s for the parallel arrangement of Fig. 14.3 is

$$\frac{1}{\lambda_s} = \frac{1}{\lambda_1} + \frac{1}{\lambda_2}$$

$$\lambda_s = \frac{\lambda_1 \lambda_2}{\lambda_1 + \lambda_2} \tag{14.2}$$

Note that the failure rate of the serial arrangement is greater than that of its parts. For example, for $\lambda = \lambda_1 = \lambda_2$, $\lambda_s = 2\lambda$. However, the failure rate of the parallel arrangement is less than that of its parts. For example, for $\lambda = \lambda_1 = \lambda_2$, $\lambda_s = \lambda/2$.

14.4 Reliability Calculation: Single Module [4]

The reliability of the single UPS module shown in Fig. 14.4 will be calculated. The module consists of a rectifier, battery, and inverter. The inverter receives dc power from the rectifier when the utility line is operating and from the battery when the utility line fails.

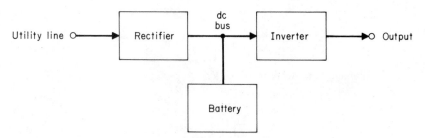

Figure 14.4 Block diagram of a single UPS module.

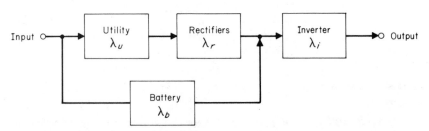

Figure 14.5 Reliability diagram of a single UPS module.

The failure rates of the components of a system are designated by λ in failures per hour or failures per million hours. A reliability diagram is drawn, as shown in Fig. 14.5. All of the elements that can individually cause the system to fail are placed in series. Elements that provide alternatives to keep the system in operation are placed in parallel. In Fig. 14.5, the utility failure rate block and the rectifier block are placed in series because the failure of either will result in loss of power from the rectifier. The battery block is placed in parallel as an alternative source to the dc bus.

The failure rate of the dc system λ_{dc} is given by

$$\frac{1}{\lambda_{dc}} = \frac{1}{\lambda_\mu + \lambda_r + 1/\lambda_b} \tag{14.1}$$

The reliability diagram of the entire module, including the output circuit breakers λ_{bk} and the power and control connections λ_c, is shown in Fig. 14.6. All of the elements are in series, so the module

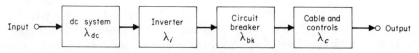

Figure 14.6 Complete reliability diagram of a single UPS module.

failure rate λ_s is given by

$$\lambda_s = \lambda_{dc} + \lambda_i + \lambda_{bk} + \lambda_c \qquad (14.2)$$

14.5 Reliability Calculation: Redundant Systems

To increase the reliability (reduce the failure rate) of such components as UPS modules, MG sets, and engine-generator sets, the components are arranged in a redundant system. Multiple components are connected in parallel and operated to share the load, as shown in Fig. 14.7. The ratings are so selected that the failure of one component and its disconnection leaves the remainder of the system with enough capacity to carry the load. The failed component is then repaired and returned to service to restore the redundancy.

The redundant system includes two types of components that can cause system failure: (1) components such that the failure of one still allows the system to operate and (2) components that are common to the system and are such that the failure of one causes the system to fail. Obviously, in the design of the system, the first type of component is protected by redundancy and the second type is built to be as reliable as possible.

The failure rate of the redundant portion of the system is dependent upon the repair time and return to service of the failed component. The repair time is defined as MTTR (mean time to repair), or repair rate $\mu = 1/\text{MTTR}$.

The failure rate λ_p of a parallel-redundant system with identical components of failure rate λ and repair rate μ is given by [6]

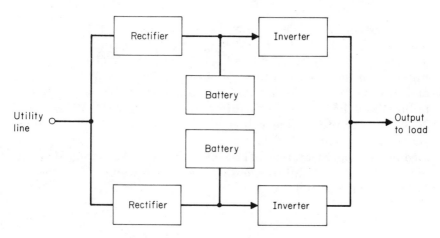

Figure 14.7 Two-module redundant system.

$$\lambda_p = \frac{k\lambda^2}{\mu} \qquad (14.3)$$

where $k = 2$ (two-component system), 6 (three-component system), 12 (four-component system).

14.6   Static Bypass Switch

An alternative way to increase the reliability of a single UPS module or MG set is to use a static bypass switch, as shown in Fig. 14.8. When the module fails, the output switch opens and the bypass switch closes to provide the load with utility power. Obviously, the voltage, frequency, and phase of the bypass line must match the load, i.e., 60-Hz line to 60-Hz load.

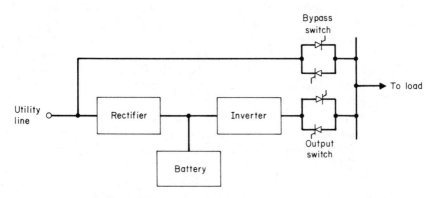

Figure 14.8   Single UPS module with static bypass switch.

The system MTBF depends on both the repair time for the failed module, when the load is being supplied direct from the utility line, and the MTBF of the utility line itself. The system MTBF is shown in Fig. 14.9 for the module of Example 14.1 as a function of utility MTBF and module MTTR. For utility line values of MTBF up to 24 h (1 failure/day), the bypass does not improve the module MTBF above its 36,350 h. For utility line values of MTBF above 8760 h (1 failure/year), the system MTBF levels off at 175,000 h, close to the two-module redundant system MTBF of 200,000 h. At a typical utility MTBF of 730 h (1 failure/month), the bypass switch improves the MTBF by a factor of 4.

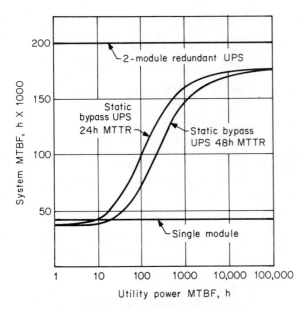

Figure 14.9 MTBF of single module with static bypass switch vs. utility power MTBF [4]. (©*1979 IEEE*).

Example 14.1 Use the failure rates in Table 14.3 to find the failure rate and MTBF for the dc system of Fig. 14.5 and for the whole module of Fig. 14.6 [4].

solution

$$\lambda_b \quad = \quad 3.57 \times 10^{-6} \text{ failures/h}$$

$$\lambda_r \quad = \quad 4.35 \times 10^{-6}$$

$$\lambda_\mu \quad = \quad 1000 \times 10^{-6}$$

$$\lambda_i \quad = \quad 20 \times 10^{-6}$$

$$\lambda_{bk} \quad = \quad 2.5 \times 10^{-6}$$

$$\lambda_c \quad = \quad 1.45 \times 10^{-6}$$

$$1/\lambda_{dc} \quad = \quad 1/(4.35 + 1000) \, 10^{-6} + (1/3.57) \times 10^{-6}$$

$$\lambda_{dc} \quad = \quad 3.56 \times 10^{-6} \text{ failures/h}$$

$$\text{MTBF}_{dc} = \quad 281{,}000 \text{ h}$$

$$\lambda_s = (3.56 + 20 + 2.5 + 1.45)10^{-6}$$
$$= 27.5 \times 10^{-6} \text{ failures/h}$$

$$\text{MTBF}_s = 36,350 \text{ h}$$

Note that the system failure rate λ_s is determined principally by the battery and the inverter.

Example 14.2   Find the failure rate and MTBF for the two-module parallel-redundant system shown in Fig. 14.7 [9]. Assume that the common portions of the system include the critical circuit breakers, power and control connections, and control commonality of Table 14.3. Use the failure rates of Table 14.3 and $\mu = 1/48$ repairs/h. Assume that the module failure rate is determined by the inverter, λ_i.

solution   The failure rate for the two parallel-redundant modules is given by

$$\lambda_p = \frac{2\lambda i^2}{\mu} = \frac{2(2 \times 10^{-5})^2}{1/48} = 0.04 \times 10^{-6}$$

The failure rate for the systems is given by

$$\lambda_s = \lambda_p + \lambda_m + \lambda_{bk} + \lambda_c$$

$$= (0.04 + 1.0 + 2.5 + 1.45)10^{-6}$$

$$= 5.0 \times 10^{-6} \text{ failures/h}$$

$$\text{MTBF}_s = 200,000 \text{ h}$$

The MTBF = 200,000 h for the two-module parallel-redundant system compares with 36,350 h for the single-module system in Example 14.1. Furthermore, the MTBF of the parallel-redundant modules alone is $1/\lambda_p = 25$ million h, compared to 50,000 h for each inverter alone. In the parallel-redundant system, the system failure rate is usually determined by the common components, not the rectifier-inverter modules.

14.7   Summary

The procedures for calculating the reliability of a given electrical system are highly developed. However, the calculations require the knowledge of failure rate of each of the components of the system. These are difficult to obtain from manufacturers or in the technical literature. As a result, calculated system reliability numbers are more useful in comparing various equipment designs or configurations for the same application than in predicting the absolute reliability of a given system.

REFERENCES

1. I. Bazovsky, *Reliability Theory and Practice,* Prentice-Hall, Englewood Cliffs, N.J., 1961.

2. M. Goldstein and P. D. Speranza, "The Quality of U.S. Commercial AC Power," *Intelec '82*, pp. 28–33.
3. "Report on Reliability Survey of Industrial Plants, Part I: Reliability of Electrical Equipment," *IEEE Committee Report, IEEE Trans. IAS,* vol. IA-10, no. 2, March/April 1974, pp. 213–235.
4. E. G. Schwarm, "Comparative Reliability of Uninterruptible Power Supply Configurations," *Intelec '79*, pp. 127–132.
5. Military Handbook 217D, "Reliability Prediction of Electronic Equipment," 1983.
6. A. Kusko and F. E. Gilmore, "Concept of a Modular Static Uninterruptible Power System," *IEEE Conf. Record of IGA 1967 Second Annual Meeting*, pp. 147–153.

Chapter 15

Installation

An emergency/standby electric power system is an independent electrical system superimposed on the utility-supplied electrical system for the facility. It is installed for safety or economic reasons. Installations for safety, as in health care facilities and public buildings, are required by codes. Installations for economic reasons are justified to avoid the cost of shutdown of data processing systems and communications facilities. Installations may also be made to peak-shave, i.e., to reduce demand. When an emergency/standby system is installed in an existing building, the arrangement of the equipment is determined by available space. When the installation is in a new building, the equipment is usually placed in a more suitable space.

15.1 Types of Equipment

An emergency/standby system includes two kinds of equipment: (1) that which is common to ordinary building electrical equipment, such as panelboards, circuit breakers, transformers, and conduits and busways, and (2) that which is "special." The special equipment includes the following:

1. UPS modules, control cabinets, static and bypass switches, alarm and monitoring panels

2. Batteries for emergency lighting, engine starting, and UPS

3. Generator sets, driven by gasoline or diesel engines or gas turbines, including fuel systems, exhaust systems, control panels, and electrical switchboards

4. Transfer switches, both as part of UPS or generator sets and standalone

5. Power distribution units in computer rooms

In addition, auxiliary equipment for heating, cooling, ventilating, and lighting is required for the emergency/standby equipment.

15.2 Equipment Locations

In new buildings, locations for UPS, batteries, and generator sets can be optimized; in existing buildings, they may be compromises. The following factors are considered in locating equipment:

1. *Accessibility.* The location must be accessible both for the original installation and for removal, if necessary, for repair or updating. There must be adequate space around the equipment to maintain it and to remove the largest part. When multiple UPS modules or generator sets are installed, maintenance work on one unit must be possible while the remaining units are running.

2. *Space and weight.* Floor loadings must be adequate to carry the weight of the equipment, particularly generator sets, batteries, and transformers. Overhead clearance must be adequate to install, maintain, and remove the equipment.

3. *Codes.* The NEC®,* NFPA, and local codes must be consulted on the location and separation of electrical equipment for emergency and legally required standby systems from the equipment for the normal electrical system. The codes must also be consulted on all aspects of generator set, battery, and transfer switch installation.

15.3 Circuits

Circuits are defined as feeders, branch circuits, and other wiring in the emergency/standby system, including conductors carrying 60-Hz power from the utility, UPS, and MG set, 415-Hz power for computers, and dc power for emergency lighting and UPS.

60-Hz power

Circuits for 60-Hz power are sized in accordance with the ampacity tables of the National Electrical Code® [1] while also taking into consideration additional conductor cross sections to limit voltage drop. The bus and panel arrangements not only must comply with the code but should be designed to facilitate maintenance. For example, systems that cannot be shut down, even for a day, must be designed with by-

---

*National Electrical Code® and NEC® are registered trademarks of the National Fire Protection Association, Inc., Quincy, Mass.

pass arrangements to permit taking individual equipment and circuits out of service for periodic testing and maintenance.

Article 700 of the NEC® provides rules for separating emergency/standby system wiring from the normal wiring as follows:

1. *Emergency systems.* Wiring from emergency source or emergency source distribution overcurrent protection to emergency loads shall be kept entirely independent of all other wiring and equipment and shall not enter the same raceway, cable box, or cabinet with other wiring. (Exceptions are cited in Art. 700-9.)

2. *Legally required standby systems.* Wiring shall be permitted to occupy the same raceways, cables, boxes and cabinets with other general wiring.

3. *Optional standby systems.* According to the NEC®, wiring shall be permitted to occupy the same raceways, cables, boxes and cabinets with other general wiring. However, to preserve the reliability required of UPS-supplied loads, the wiring should be separated from general wiring to prevent noise coupling, and to permit maintenance mishaps and identification.

Circuits carrying 415-Hz power from UPS or MG sets require special design, as discussed in the following subsections.

415-Hz power

Design of a 415-Hz system requires the same procedures of selecting conductors and conduits for voltage drop and ampacity as the design of a conventional 60-Hz system [2]. At 415 Hz, the inductive reactance and ac effective resistance have significant effects on the circuit. The voltage drop per unit length is greater. The magnetic field of the current increases the losses in the conductors, conduits, and adjacent ferromagnetic materials.

The effect of the 415-Hz current on the losses is demonstrated in Fig. 15.1. Within the conductor itself, the magnetic field forces the current to the periphery of the conductor, the so-called skin effect. This increases the effective resistance from the dc value represented by the entire cross section to the ac value represented by the cross section of the "skin."

The effect of the conduit depends upon whether the conduit is ferromagnetic, metallic, or nonmetallic. In a ferromagnetic conduit, the inner wall carries an ac magnetic field produced by the currents of conductors lying adjacent to the wall and any net unbalance current of the three conductors. This ac magnetic field produces eddy current losses which heat the conduit and the conductors and add to the effec-

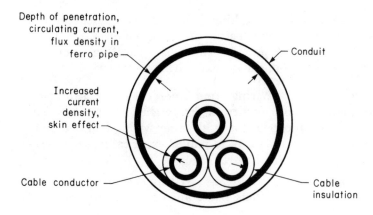

Depth of penetration,
circulating current,
flux density in
ferro pipe

Conduit

Increased
current
density,
skin effect

Cable conductor

Cable
insulation

Figure 15.1 Section showing three conductors operating at 415 Hz in a metallic conduit [2].

tive conductor ac resistance. In a metallic conduit such as aluminum, the conduit will carry longitudinal eddy currents proportional to the conductor current, which will produce heating. In a nonmetallic conduit, there will be no heating effects; however, the magnetic field of the conductors will not be contained by the conduit and may cause external problems.

The increase in frequency from 60 to 415 Hz increases the inductive reactance proportionally and thereby increases the voltage drop by seven times. The reactance itself can be reduced by decreasing the distance between conductors, i.e., using thinner insulation. When parallel feeders are used, the reactance can be lowered if the individual conductors of the same phase are separated as much as possible and if the distances between conductors of different phases are kept small. This is the principle on which the construction of "balanced voltage" bus is based. The presence of magnetic materials increases the flux density in the area of the conductors and thus increases the reactance as well. The flux density and reactance can be reduced by the use of nonmetallic conduit, aluminum conduit, or aluminum-armored cable.

The conductors and conduits of the 415-Hz circuits can be designed by following the manufacturers' specific recommendations or by working from tables of conductor parameters. An example of the recommendation for the 415-Hz circuits for an IBM 3081 processor unit which requires 40 kVA 208-V three-phase power is given in Table 15.1 [3]. Tables of conductor parameters at 415 Hz are given by Lauderbaugh and Rhoades [2].

The ampacity of a conductor at 415 Hz, relative to 60 Hz, falls off quite rapidly as the conductor size is increased. For example, a No. 4/0 THW copper conductor installed in steel conduit and operating at 75 °C

TABLE 15.1   Circuit Selection Guide for 415-Hz, 208-V Power for IBM 3081 Processor Unit [3]

Copper wire size*,§	Conduit		Maximum run lengths‡ for rigid aluminum conduit, m (ft)¶
	Quantity†	Size, mm (in)	
2 AWG	1	40 (1½)	42 (139)
1/0 AWG	1	50 (2)	49 (164)
2/0 AWG	1	50 (2)	53 (177)
2 parallel AWG	2	40 (1½)	85 (279)
1/0 parallel AWG	2	50 (2)	100 (329)
2/0 parallel AWG	2	50 (2)	107 (351)

*Single runs with conductors smaller than 2 AWG should not be used. Three-conductor jacketed cable, with ground conductor, is acceptable.

†Conduit quantity refers to the number of conduits recommended; each conduit contains all three phase conductors and an insulated equipment-grounding conductor in the wire size shown.

‡Maximum run lengths assume maximum voltage drops at 4% with 100-A load and a power factor of 0.766 lagging.

§Feeders using other than single 2 AWG conductors require that each phase and ground conductor (or parallel conductor) be terminated in a junction box in a single 2 AWG conductor for connection to the Russell & Stoll receptacle. (The Russell & Stoll receptacle does not accommodate conductors larger than 1 AWG.)

¶Lengths are rounded to the nearest meter. Local and national wiring codes must be followed.

at 415 Hz has an ampacity equal to only 76 percent of its 60-Hz rating. Increasing the size of the conductor to decrease the voltage drop is not effective. Nothing can be gained by increasing the size of circuit conductors to more than that required by the current and based on temperature rise.

The proper solution to a voltage-drop problem is to subdivide the circuit into parallel conductors and arrange the conductors to provide the least reactance. The selection of the number and size of the parallel conductors requires an understanding of the costs to determine the balance between improved performance and increased cost. It is suggested that the designer tabulate the impedances, voltage drops, and costs for various combinations to permit the most economical choice.

The voltage drop caused by the reactance of a 415-Hz circuit can be reduced by one of at least three methods: (1) For a single circuit supplied by a dedicated UPS module or MG set, the signal for the voltage regulator can be sensed at the load bus instead of at the UPS or MG set terminals. (2) The signal for the voltage regulator can be sensed at the terminals of the UPS or MG set, and a second signal proportional to the circuit impedance voltage drop can be subtracted to simulate the voltage of the load end of the circuit [3]. The inductive reactance of the circuit can be compensated by capacitors to minimize the total voltage drop. A line-drop compensator (LDC) of this type is shown in Fig. 15.2 [4].

Figure 15.2   Line drop compensator for 415-Hz circuit [4].

15.4   Protection

Although the protection equipment for an emergency/standby system must be designed to prevent the destruction of the system from overloads and faults, it must also ensure that the system fulfills its function of supplying critical loads in an emergency. The design of the protection equipment differs from that of the normal utility-supplied 60-Hz facility in the following ways:

1. *Fault current may be low.*   The fault current may be delivered by engine-generator sets having lower kVA and higher impedances than the normal transformers. It may come from static UPS which are current-limited in overload. It may come from ferroresonant regulating transformers that limit output current to about 150 percent of rated. Feeders and branch circuits may have to be subdivided by using smaller fuses or circuit breakers that can clear branch circuit faults under the limited fault currents delivered by the emergency/standby source.

2. *Protection system must not trip on switching.*   Transfer switching, module switching, and engine-generator start-up occur at the time of utility line failure and restoration and when equipment fails. Breaker trip settings and fuse ratings must be adequate for the currents of restarting motors, reenergizing rectifiers with capacitor filters, picking up discharged battery banks, and supplying magnetizing

current to energizing transformers. In the design of each system, the sequencing of loads must be so scheduled that protective equipment will not trip during an emergency condition.

3. *Engine-generator, MG, and UPS protection must be coordinated with branch circuit and feeder protection.* The manufacturer provides internal protection for the equipment, e.g., overload relays, ground fault detection, and overtemperature detection. The circuit breakers and fuses of the circuits that extend into and out of the equipment must be coordinated with the internal protection of the equipment. For example, a fault in a branch load circuit must not trip an entire UPS.

4. *Battery system protection requires a unique design.* DC fault currents rise exponentially with time, unlike ac fault currents, which pass through zero at every half cycle of time. DC fuses and circuit breakers do not clear faults at the current zeroes at which ac protective devices operate. DC devices must be certified for specific dc fault duty. Battery system fault currents have high rates of rise and high ultimate fault currents. Battery manufacturers sometimes design the intercell links to serve as a backup to the dc protective devices.

Article 700 of the NEC® [1] includes requirements for overcurrent and ground fault protection of emergency/standby systems.

15.5   Generator Protection

Neutral grounding and ground fault protection systems are needed for engine-generator sets, because [5,6]:

1. Large sets which supply four-wire emergency service to factories, health care facilities, office buildings, and elevators—and standby service to data processing centers—must now comply with the neutral grounding requirements of the National Electrical Code® (NEC®).

2. Cogeneration and peak-shaving sets which are being installed to operate in parallel with the utility must comply with the utility requirements for neutral grounding protection.

The generator neutral can be grounded through an impedance that ranges from no ground (infinity) to a solid ground (zero). In order to sense ground faults within the generator windings or within the electrical distribution system, and to trip the generator, the protection system must be matched to the neutral grounding method.

A ground fault is the most common type of electrical failure. It occurs whenever there is a breakdown in the insulation of a cable, winding, or any other electrical component. The ground fault current often is not high enough to trip overcurrent devices. However, if it goes undetected and isolated, it escalates to the more serious three-phase fault which leads to fire and serious damage. Therefore, a separate ground fault protection scheme is required. The best method of grounding the neutral and protecting against the resultant ground fault current depends on how the engine-generator is going to be applied.

Engine-generator sets of up to 3000 kW 208Y/120 V or 480Y/277 V are used to supply loads continuously or in the standby mode. Figure 5.12 shows a standby generator connected to a low-voltage three-phase four-wire system by means of a transfer switch. A low-impedance path to the generator neutral is needed here to supply an unbalanced three- or single-phase load. The NEC® requires that this neutral conductor be grounded, since it is used as a service conductor. See Ref. 5.

To meet the double requirements, the neutral of the generator must be either solidly grounded or grounded through a low reactance. Per NEMA standards, the generator windings are braced to withstand a solid three-phase fault but not a line-to-ground fault. To limit the ground-fault current to the three-phase fault level, the neutral is grounded through a low (ohmic) reactance. Normal four-wire operation is possible because the low reactance adds little to the voltage drop for one-phase loads. An air core reactor is used to avoid saturation by the fault current. In the event of a ground fault, it prevents overvoltages from occurring in the unfaulted phases.

As shown in Fig. 5.12, ground fault protection is provided by the overcurrent relay 51N in the neutral and the differential relays 87 in the winding circuits of the generator. For all ground faults in the generator windings the differential relays trip the generator breaker without delay; the relay 51N acts as a backup to the differential relays. For external ground faults the overcurrent relay 51 trips the breaker. The pickup of this relay is set just above the expected maximum neutral current produced by one-phase or unbalanced loads.

15.6 Grounding

Emergency/standby systems fall into two broad categories with respect to grounding: (1) a normal utility system and (2) a separately derived system. A separately derived system is one whose power is derived from generator, transformer, or converter windings and which has no direct electrical connection, including a solidly connected grounded circuit conductor, to supply conductors originating in another system [1]. Furthermore, two types of conductors are used in grounding:

1. *Grounded conductor.*   A system or circuit conductor that is intentionally grounded, e.g., an insulated neutral conductor (white wire) that is grounded at the service entrance.

2. *Grounding conductor.*   A conductor used to connect equipment or the grounded circuit of a wiring system to a grounding electrode (grounded metal structure or water pipe), e.g., an insulated green wire or bare conductor.

Electrical equipment or wiring systems are grounded for one or more of the following reasons:

1. *Provide a path for fault current.*   When a live conductor makes contact with the metal parts or housing of the equipment, the fault current returns to the distribution panel by the grounding conductor and causes the fuse or circuit breakers to operate.

2. *Ensure the safety of personnel.*   For the fault described in item 1, grounding ensures the safety of a person in contact with the equipment by (1) causing the electrical supply to be interrupted quickly and (2) maintaining the equipment housing at a potential less than that of the live conductor.

3. *Reduce static charge.*   The ground provides a path for static electric charge to leak off equipment, such as printers, terminals, processors, disk drives, to ensure that (1) operation is not compromised, (2) semiconductors are not damaged, (3) sparks do not cause fires or explosions.

4. *Reduce electrical signal noise.*   The ground provides a path for noise signals to be bypassed from cable shields, filters, and circuits to ensure operation.

5. *Reduce electric field coupling.*   The grounding of metal parts of signal-level equipment reduces the unwanted coupling between different parts of the circuits.

6. *Provide a path for currents caused by lightning.*   The grounding conductors allow the currents to drain to ground while the respective potentials of equipment are limited to nondangerous and nondestructive values.

The National Electrical Code® (NEC®), Art. 250 [1], describes the safety aspects of grounding. FIPS Pub. 94, Chap. 3 [7], describes grounding for data processing systems. Denny [8] describes grounding for the control of EMI.

The grounding of the portions of the emergency/standby system that are supplied from the normal utility system follows the rules and practices given in the NEC®. The grounding of the portions that are defined as "separately derived," i.e., from generators, isolation transformers, and a UPS, must also follow the NEC®. The system grounded conductor (neutral) is intentionally grounded at only one point—at the source and ahead of any system-disconnecting means or overcurrent device. The grounding of equipment in a computer room requires coordination of NEC® safety-derived rules and requirements for correct operation of signal-handling circuits.

The first lesson to be learned in grounding is that the electrical system operates with a separate equipment grounding wire and a neutral wire. The grounding wire is colored green. It is earth-grounded at the service entrance at distribution panels and at other places. Its primary function is to provide an electrical path for fault current to return to the panel while it holds down the potential of faulted equipment.

The neutral wire is classed as a grounded conductor and is colored white in one-phase and three-phase systems. Its function is to provide an electrical path for load current to return to the panel. In three-wire one-phase or four-wire three-phase systems, the neutral carries the unbalance current. In three-phase systems, the neutral may also carry 3d-harmonic current from nonlinear loads. The neutral also is earth-grounded at the service entrance, but nowhere else. However, if the secondary of an isolation transformer or a UPS becomes the source, then the neutral must be earth-grounded at the separately derived source.

The procedure for designing a grounding system for an assembly of equipment such as computers and peripherals includes the following:

1. *Separate the signal and equipment grounds in each equipment.* Typically, the manufacturer connects the grounds together within the equipment and brings them out to one terminal. Also be sure that the shields of interconnecting signal cables are open at one end to ensure that the shields do not connect together the grounds of individual equipment.

2. *Form single-point grounds.* Bring all of the equipment grounds to one ground bus and all signal grounds to the same bus. Earth-ground the ground bus.

3. *Equipment-ground sizes.* The equipment-grounding cables must be sized to carry the maximum ground fault current until the protection devices clear the fault.

The stator winding of a standby generator set is grounded to meet the requirements of the set as a source of power. For four-wire service (with neutral), the neutral can be solidly grounded or grounded through a reactor, as shown in Fig. 5.12. The impedance of the reactor is selected to limit the line-to-ground fault current to the three-phase fault current. The neutral of the generator also can be grounded through a high impedance, which will limit the stress on the winding of all line-to-ground faults.

15.7 Maintenance

Emergency/standby systems require equipment beyond that required for a conventional building electrical system, e.g., UPSs, batteries, engine-generator sets, and transfer switches. This equipment requires organized maintenance to ensure the necessary reliability. Equipment failures must be anticipated. The issues discussed in the following subsections must be considered.

Maintenance procedure

Maintenance can be conducted by building personnel or by contract personnel. Building personnel must be trained for the equipment and be readily available when a failure occurs. However, sufficient personnel must be trained to cover personnel absent during vacations and sick leave. Contract personnel are trained for their own equipment. Since they are maintaining the same type of equipment continuously, they would be expected to be more skillful at maintaining it. They are also capable of making field changes to update the equipment, and they have access to the factory for backup. To be fully effective, they must be available promptly if a failure that results in the shutdown of critical equipment occurs.

Preventive maintenance

Preventive maintenance is a blessing and a curse. It must be carried out to find incipient failure conditions, such as loose connections, potentially failed parts, specific settings. Typically, intervals are either every quarter or half year. However, equipment failures after preventive maintenance are not unusual. Connections may not be returned; parts or tools may be left inside cabinets; reenergizing procedure may not be followed. Tight supervision and checking must be done during preventive maintenance by building or contract personnel. Accurate records of what was found and what was done should be kept.

Documentation

To ensure effective maintenance, personnel must maintain all of the documentation on the equipment including the following:

1. As-built electrical and site drawings of the equipment
2. All manuals received from the vendors, updated with all modifications
3. Maintenance records
4. Any special procedures generated by the vendors or the maintenance supervisor
5. Names, telephone numbers, and addresses of all personnel who may be required in the event of emergency failure conditions
6. Original specifications, contracts, and purchase orders
7. Spare parts inventory list

Spare parts

Spare parts must be maintained at the site to handle predictable failures, e.g., fuses, semiconductors, relays, and circuit boards. The inventory can be maintained by building personnel or contract personnel. When a spare part is used, it should be replaced. When a component is repaired at the factory and returned as a spare part, e.g., a circuit board, it should be tested in the equipment before being placed in the spare parts inventory. The spare parts inventory should be checked as part of the preventive maintenance.

15.8 Summary

The emergency/standby electric power system is an independent electrical system superimposed on the utility-supplied electrical system for the facility. Installation involves equipment that is common to ordinary electrical facilities and equipment that is special, such as static UPS, batteries, MG sets, engine-generators, and transfer switches. Electric circuits must meet NEC® requirements and special requirements for dc battery circuits, 415-Hz computer circuits, and grounding. Installation standards must ensure a high order of reliability compared to the utility supply. Equipment must be maintainable, accessible, well documented, and safe.

REFERENCES

1. ANSI/NFPA 70-1987, "National Electrical Code®."
2. M. E. Lauderbaugh and B. M. Rhoades, "A Guide to 400-Hz Power Distribution," *Actual Specifying Engineer,* February 1972, pp. 73–83.
3. "IBM Guide to 400-Hz Power Requirements," *GC 22-7070-1, File No. S 370-15,* 1981.
4. R. J. Lawrie, "Designing and Installing a Power System for a Large Computer Center," *EC&M,* January 1986, pp. 76–83.

5. A. Kusko and S. M. Peeran, "Neutral Grounding and Protection for Energy Saving Generator Sets," *Engineer's Digest,* April 1987, p. 64.
6. A. Kusko and S. M. Peeran, "Minimize Cost of Engine/Generator Protection with Static Relays," *Power,* May 1987, pp. 35–38.
7. "Guideline on Electrical Power for ADP Installations," *FIPS Pub. 94,* U.S. Dept. of Commerce/National Bureau of Standards, Sept. 21, 1983.
8. H. W. Denny, "Grounding for the Control of EMI," Don White Consultants, Inc., Gainesville, Va., 1983.

16

Procurement

Each emergency/standby electric power system installation is unique. The objective of the procurement process is for the end user (customer) to obtain the most reliable system for the least lifetime cost. The procurement process involves at least four parties: the end user, the consulting engineer, the equipment vendors, and the installation contractor. The process starts with the formulation of the requirement and ends with the system properly supplying power to the designated loads.

Emergency/standby electric power systems are installed under two conditions: (1) as part of a newly constructed facility or (2) as a modification of or addition to an existing facility. The procurement process is the same in either case.

In this chapter, we define "component" as an integral piece of equipment, e.g., engine-generator set, transfer switch, or UPS module. We define "system" as the complete assembly of the components to form the emergency/standby system.

16.1 Procurement Process

The first step in the procurement process is for the end user, with the help of a consulting engineer, to determine the system requirements, including:

1. Function of the emergency/standby system

2. Loads to be served

3. Space available for equipment

4. Dollar and time budgets

5. Anticipated growth of load

6. Reliability level

7. Operating and maintenance responsibility

The tasks following the determination of system requirements include:

1. Engineering

2. Preparing specifications

3. Evaluating bids and vendors

4. Purchasing the components

5. Testing components at the factory and site

6. Constructing and integrating the system

7. Training operating personnel

8. Establishing maintenance program

These tasks will be treated in the following sections.

16.2 Engineering

Engineering for an emergency/standby system should be done by an electrical engineering firm that has experience in such systems. After reviewing the system requirements, the firm will do the following:

1. *Analyze the electrical loads.* Classify them into levels of required power reliability in accordance with the appropriate codes, or by function for noncode installations such as the function of data processing. Formulate the method of system operation under all contingencies.

2. *Prepare the design of the system.* Include the electrical design, protection, physical design, and grounding system.

3. *Prepare the specifications.* Write specifications for each component that must be purchased and for the overall construction.

Assist the end user with the remaining steps of the procurement process.

16.3 Specifications

A clearly written set of specifications for the components and construction is crucial for the realization of a properly operating emergency/standby system. The specifications are the heart of the "contract" between the vendor and the end user to purchase the equipment. The specifications should not make any requirements that cannot be verified by inspection, calculation, or testing. The specifications must spell out the conditions for the acceptance of the equipment so that the vendor can get paid and the contract be concluded. Specifications are costly and time consuming to write. The following sources can be used:

1. Sample specifications prepared by the vendor

2. Specifications prepared by the consulting engineer or for the end user for previous projects

3. Specifications written from scratch but using the format set by the end user or other organization

An example of the outline of a vendor's sample specification for a parallel multiple module uninterruptible power system is given in Table 16.1 [1].

16.4 Purchasing

Managing of the purchasing of the components for an emergency/standby system can be done by the end user, the overall contractor, or the consulting engineer [2,3]. Components such as UPS modules, transfer switches, and batteries are not found in conventional electrical systems; they may be unfamiliar to the purchaser and not identifiable by standard names in the electrical industry. In any case, the purchasing process includes the following steps:

1. *Prepare bidders' lists.* In addition to bidders supplied by the consulting engineer, trade journals publish special issues on emergency/standby power systems that include the names of possible bidders.

2. *Issue specifications.* Include all selected bidders for each component. Respond to questions by possible bidders. Issue classifications or corrections to specifications, as necessary. If an existing facility is to be expanded, the end user may want to expand by using the same manufacturers.

3. *Evaluate bids.* Reduce the number of bids to two or three for each component based on price, delivery, compliance with specification, and experience.

4. *Evaluate vendors.* Visit vendors' plants to inspect components

TABLE 16.1   Outline of Specification [1]: Parallel Multiple-Module Uninterruptible Power System

1.0 General
   1.1 Submittals
   1.2 Installation

2.0 Description and operation
   2.1 Definitions
   2.2 Configuration

3.0 UPS requirements
   3.1 Rating
   3.2 Electrical characteristics
   3.3 Environmental conditions
   3.4 Audible noise
   3.5 Grounding
   3.6 Efficiency

4.0 UPS module
   4.1 Rectifier/charger unit
   4.2 Inverter unit
   4.3 Protection
   4.4 Module control section
   4.5 UPS module battery circuit breaker

5.0 UPS system control cabinet
   5.1 Control and monitoring section
   5.2 Power section

6.0 Storage battery

7.0 Remote alarm panel

8.0 Self-diagnostic circuitry

9.0 Equipment details
   9.1 Wiring
   9.2 Construction and mounting
   9.3 Ventilation

and facilities at first hand; visit sites where the vendors' equipment is installed; and determine the experience of others with the components.

5. *Select vendors.*   Work out delivery schedules, changes from the specifications, and terms of acceptance and payment. Place orders.

6. *Witness tests.*   Witness factory and site tests of the equipment as described in Sec. 16.5.

7. *Organize maintenance.*   Make arrangements for maintenance as described in Sec. 16.6.

In the purchasing process, it is important for the end user and the vendors to complete the process in such a manner that the vendors will, in the future, assist in engineering and field changes, getting spare parts, and making emergency repairs.

16.5   Testing

The proof that good equipment has been purchased is demonstrated by testing in the factory and at the site. The test procedures must be spelled out in the specifications. The following is a typical sequence of tests for a component of the emergency/standby system:

1. *Factory tests.*   The vendor submits a test plan to the end user. After approval of the plan, the end user, or his representative, witnesses the tests. If the tests are satisfactory, the equipment is approved for shipment. If the tests fail, the equipment must be modified and the tests must be repeated. Frequently, complete equipment is not assembled at the factory or the factory may not have the electrical or other facilities for the test. Only part of the equipment is tested at the factory.

2. *Site test.*   The tests made in the factory are repeated at the site by using all of the assembled on-site equipment, such as batteries, engine-generator sets, and switchgear. The equipment is usually tested with dummy loads, which may be built into the site or brought in for the tests.

3. *Site tests with actual loads.*   The actual loads, e.g., data processing equipment, are picked up by the installed equipment. All of the emergency and standby sequences, such as engine-generator set starting, transfer switch operation, UPS transfers between utility supply and battery supply, are tested.

16.6   Operation and Maintenance

Emergency/standby systems are so designed that practically no manned operation is required. When utility power fails, the system is supposed to take over automatically and supply the electrical loads in the prescribed manner. However, the systems do require periodic testing, preventive maintenance, and repairs. Most failures of emergency/standby systems are the result of inadequate maintenance.

When the system is being designed, constructed, and placed in operation, the following questions must be resolved for each component of the systems.

1. *Who will do the maintenance?*   It can be done by trained on-site personnel, by the vendors' organization, or by an independent contractor.

2. *What spare parts are required?*   They can be provided and maintained by the end user or by the maintenance organization. The vendor also can provide a backup supply. An orderly process is required to keep the stock up to date.

3. *What will be the preventive maintenance program?*   Maintenance can be done annually, semiannually, or at intervals recommended by the vendor.

4. *Who will keep the maintenance records and update the drawings and instruction manuals?*   All of the material should be the responsibility of one person or office and be kept in a central place at the site.

16.7   Summary

The procurement process starts with the system considerations and ends with the emergency/standby system in troublefree operation. All of the steps in the process, i.e., engineering, specifying, purchasing, installing, testing, and organizing the maintenance, require good management. The technology of the equipment is much advanced over that in the conventional electrical system.

REFERENCES

1. "Specification for a Parallel Multiple Module Uninterruptible Power System," Emerson Electric Co., Santa Ana, Calif.
2. A. Kusko and T. Knutrud, "Purchasing and Installing Solid-State Rectifier Sets," *Electrical Construction and Maintenance*, August 1975.
3. A. Kusko and T. Knutrud, "Specifying Uninterruptible Power Supplies," *Electrical Construction and Maintenance*, February 1974.

17

Cost/Benefit Analysis

Cost/benefit analysis is a normal step in the design of any system and the selection of its components. For emergency/standby systems, cost/benefit analysis can be conducted for the following typical situations:

1. For a system required by code, when a standby/emergency system is required, but the actual design has not been specified.
2. For a critical load which is disturbed by utility power failures, when the decision to install UPS is considered on economic grounds.
3. For a UPS, when the use of rotary or static equipment and the number of units must be selected.

Although the theory of cost/benefit analysis is sound, its execution for emergency/standby systems is difficult. Typical questions for each facility include:

1. What are the predicted frequency and duration of utility power failures that prompt the need for emergency/standby systems?
2. What is the benefit of maintaining power to the loads during the utility power failures?
3. What is the reliability of the emergency/standby systems?

Various methods of conducting those analyses can be used. They include:

1. Initial (capital) costs of various approaches to satisfy a specific need, such as meeting a code requirement.
2. Annual costs of various emergency/standby systems, including in-

terest on capital, maintenance, fuel and electric energy, to be compared with the cost penalty (benefit) of utility power loss for the equipment to be protected.

3. Percent worth of the costs and the penalty (benefit) of the annual costs of item 2.

Examples of these methods will be presented in Sec. 17.4.

17.1   Frequency of Utility Power Failure

The frequency and duration of utility power failures should be determined at a site before a standby/emergency system is designed. The pattern will be used to determine battery capacity for code-required systems and UPSs, fuel capacity for engine-generator sets, and whether engine-generators are required to back up battery storage systems. Either a site study can be made or published data from other sites can be used.

Data on the frequency and duration of utility power failures are dependent on the method of utility supply, the geographical location, the particular utility, and many other factors. For example, a facility located in a city and supplied from a utility low-voltage underground street network can expect an extended outage ranging from hours to days about once every 10 years. A facility located in a suburban or semirural area and supplied by an overhead primary feeder from a distribution substation can expect a voltage disturbance to disturb a computer center at least once per month and a disturbance lasting minutes to hours at least once per year.

Samples of data on utility performance are given in Sec. 14.2. An additional sample is given in Fig. 17.1 [1]. Data on every utility-caused customer outage on the Pennsylvania Power & Light Co. system were collected. The average annual frequency of interruption of primary feeders is shown in Fig. 17.1a; the average interruption duration as a function of percent of primary feeders is shown in part b.

17.2   Cost of Utility Failures

As expected, the cost of a utility power failure in a data processing center, manufacturing plant, or other facility is highly dependent upon the type of equipment interrupted and both the size and restart time of the facility. The cost can be obtained from the records of previous utility power failures or by calculating the costs for short, medium, and long failures.

Samples of published data on the cost of utility failures for a group of facilities are given as follows:

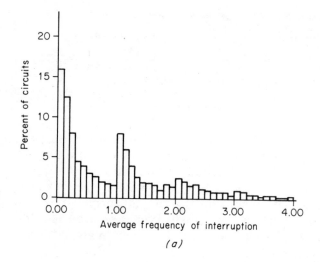

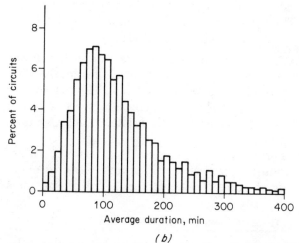

Figure 17.1 Primary feeder outage data collected on Pennsylvania Power & Light Co. system [1]. (*a*) Average annual interruption frequency showing percent of circuits as a function of number of interruptions. (*b*) Annual interruption duration showing percent of circuits as a function of average duration.

1. IEEE Committee Report, "Reliability Survey of Industrial Plants, Part II: Cost of Power Outages, Plant Restart Time, Critical Service Loss Duration Time, and Type of Loads Lost Versus Time of Power Outages," 1974 [2]. Data were collected from 30 companies reporting on 68 plants in nine industries in the United States and Canada.

TABLE 17.1   Median Cost of Power Outages for Industrial Plants in the United States and Canada, 1972 [2] (©1974, IEEE)

Plants	Cost of outage
All	$0.69 per kW + $0.83 per kWh
T 1000 kW max. demand	$0.32 per kW + $0.36 per kWh
S 1000 kW max. demand	$3.68 per kW + $4.42 per kWh

Table 17.1 gives a summary of the median cost of power outages found in the study. The first term is based on the demand when the plant is operating at its design capacity, in kilowatts. It represents the extra expense because of the failure. The second term is based on the undelivered kilowatthours during the downtime. It represents the value of downtime, i.e., lost production, measured as sales price of product not made, less expenses for labor, material, etc., saved.

The typical loads lost vs. times of power outage in the study are shown in Table 17.2. The computers were either protected by UPS or were older types that were doing batch computation.

2. Billington, Wacker, and Subramaniam [3]. Data on utility power interruption cost were collected by questionnaire from nearly 1000 commercial customers of the Manitoba Hydro system. The cost factors and the breakdown of components by standby category are given in Table 17.3. Some respondents have more than one standby system, e.g., a UPS plus an engine-generator.

The outage costs are normalized in two ways: (1) in terms of dollars per kilowatthour of annual energy consumption and (2) in terms of dollars per kilowatthour of annual peak demand. Unlike the costs

TABLE 17.2   Loads Lost vs. Times of Power Outage, 1972 [2] (©1974, IEEE)

Type of load	For equipment failures 1 cycle or less in duration, %			For equipment failures between 1 and 10 cycles in duration, %			For equipment failures 10 cycles or more in duration, %		
	Yes	No	Not known	Yes	No	Not known	Yes	No	Not known
Computer	0	0	0	4	96	0	9	91	0
Motor	0	0	0	33	67	0	67	33	0
Lighting	0	0	0	22	78	0	38	61	2
Solenoid	0	0	0	22	74	4	25	66	9
Other	0	0	0	7	15	78	25	62	13
Average plant outage duration	0.0 h			1.39 h			22.6 h		

Only nonzero data were used in computing the average plant outage duration.

TABLE 17.3   Factors in Determining Cost Estimates
of 979 Power Outages [3] (©1986-IEEE)

Cost Factors Used in Commercial Cost Estimates

Paid staff unable to work
Loss of sales
Start-up costs
Spoilage of food
Damage to equipment/supplies
Other costs or effects

Breakdown of Respondents into Categories	
No standby equipment	87%
Battery standby system	10
Engine-driven system	4
Other system	1

given by the IEEE Committee in Table 17.1, these outage costs are not
additive. The costs for the December through March period are given
in Table 17.4 for the category of no-standby systems. The costs are in
1980 Canadian dollars. The dollars per kilowatt annual peak demand
for all three categories as a function of interruption duration are
shown in Fig. 17.2.

TABLE 17.4   Factors in Determining
Cost Estimates of Power Outages,
1980 Canadian Dollars [3]
(©1986-IEEE)

Duration	Cost
1 min	$ 63.44
20 min	269.91
1 h	668.96
4 h	2451.35
8 h	$ 6589.78
1 min	.00034529 $/kWh
20 min	.00242090
1 h	.00556600
4 h	.02201700
8 h	.06562500
1 min	1.04 $/kW
20 min	5.91
1 h	15.65
4 h	57.69
8 h	148.93

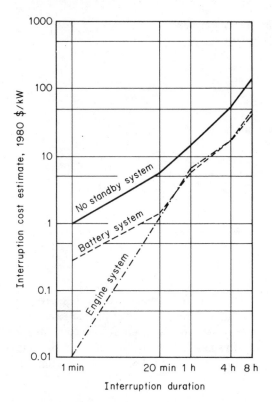

Figure 17.2 Commercial customers' interruption costs grouped by standby category, 1980 Canadian dollars [3]. (©*1986 IEEE*)

Example 17.1   *Static vs. rotary UPS: initial cost basis.* A load of 900 kW must be supplied. Two types of equipment are available: a rotary UPS with 500 kW/unit at $500/kW and a static UPS with 400 kW/module at $450/kW. Compare the initial costs for (*a*) a nonredundant system and (*b*) a one-unit redundant system:

solution   (*a*) *Nonredundant system:*

      Rotary   2 × 500 kW units at $500/kW = $500,000

      Static   3 × 400 kW modules at $450/kW = $540,000

(*b*) *One-unit redundant system:*

      Rotary   3 × 500 kW units at $500/kW = $750,000

      Static   4 × 400 kW modules at $450/kW = $720,000

The initial costs of rotary units are lower than those of the static units for the nonredundant system but higher than those for the one-unit redundant system.

Example 17.2   *Static vs. rotary UPS: annual cost basis.* Compare the additional annual costs for operating the UPS of Example 17.1 for a 900-kW load for 8760 h/year. In addition to the ratings and costs given in Example 17.1, include the following:

Interest plus amortization	10 percent
Energy	$ 0.10/kWh
Rotary maintenance	$ 1000/year unit
Rotary efficiency	0.80
Static maintenance	$ 2000/year unit
Static efficiency	0.88

Consider (*a*) a nonredundant system and (*b*) a one-unit redundant system.

solution   (*a*) *Nonredundant system.* If the UPS is the rotary type:

Interest	0.10 × $500,000 =	$ 50,000/year
Maintenance	2 × $1,000 =	2,000
Energy (losses)	0.2 × $0.10/kWh × 900	
	kW × 8760 h =	157,680
Additional cost		$209,680/year

If the UPS is the static type:

Interest	0.10 × $540,000 =	$ 54,000/year
Maintenance	3 × $2,000 =	6,000
Energy (losses)	0.12 × $0.10/kWh × 900	
	kW × 8760 h =	94,608
Additional cost		$154,608/year

(*b*) *One-unit redundant system.* If the UPS is the rotary type:

Interest	0.10 × $750,000 =	$ 75,000/year
Maintenance	3 × $1,000 =	3,000
Energy (losses)	Same as in (*a*) =	157,680
Additional cost		$ 235,680/year

If the UPS is the static type:

Interest	0.10 × $720,000 =	$ 72,000/year
Maintenance	4 × $2,000 =	8,000
Energy (losses)	Same as in (*a*) =	94,608
Additional cost		$174,608/year

The additional annual costs for the static UPS units are lower than for the rotary UPS units because of the higher efficiency and the relatively large energy costs compared to the other costs. Further reductions of additional annual costs can probably be achieved by considering units of higher efficiency but at higher initial costs. In the case of rotary UPS, higher efficiency is achieved through increased size. In the case of the static UPS, higher efficiency can be achieved through improved circuits and increased size.

Example 17.3   Cost of outage vs. cost of UPS. Compare the annual cost of outages with the cost of a UPS capable of operating the critical loads through the outages. The outages are estimated at 5 per year at a cost of $10,000/outage, or a total of $50,000/year. Operation and cost data for the UPS are the following:

UPS rating	500 kW
Operation	400 kW
12 h/day at 400-kW load	0.88 eff.
12 h/day at 200-kW load	0.88 eff.
Interest plus amortization	10 percent
Initial cost	$225,000
Maintenance	$2,000/y
Energy	$0.10/kW

solution   The additional energy required for the UPS over and above the load without the UPS is given by the UPS losses:

$$\text{Losses} = 0.12 \times 4380 \text{ h/y} \times 400 \text{ kW} + 0.20$$
$$\times\ 4380 \text{ h/y} \times 200 \text{ kW} = 385{,}440 \text{ kWh/year}$$

The annual additional cost of the UPS is given by

Interest	$0.10 \times \$225{,}000$ =	$22,500/year
Maintenance		2,000
Energy	385,440	
	kWh/year × $0.10	
	kWh =	38,544
Additional cost		$63,044/year

The additional annual cost of the UPS of $63,044 exceeds the cost of five outages per year of $50,000.

17.3 Summary

Cost/benefit analysis is carried out as a step in the engineering of all emergency/standby systems to ensure that the design satisfies the requirements at the lowest cost. Both the frequency/outage duration of the utility power and the resultant cost to the customer are difficult to quantify. Published data and methods for normalizing the data are presented in this chapter. Cost data and examples also are given.

REFERENCES

1. R. W. Filipovits and R. H. Osborn, "Outage Analysis Improves System Reliability, Transmission and Distribution," August 1986, pp. 32–36.
2. IEEE Committee Report, "Report on Reliability Survey of Industrial Plants, Part II: Cost of Power Outages, Plant Restart Time, Critical Service Loss Duration Time, and Type of Loads Lost Versus Time of Power Outages," *IEEE Trans. on Ind. Appl.*, vol. IA-10, no. 2, March/April 1974, pp. 236–241.
3. R. Billington, G. Wacker, and R. K. Subramanian, "Factors Affecting the Development of a Commercial Customer Damage Function," *IEEE Trans. Power Syst.*, vol. PWRS-1, no. 4, November 1986, pp. 28–33.

Part

5

Codes and Standards

Chapter 18

Codes Governing Emergency/Standby Electric Power Systems

A wide array of codes governs the broad field of emergency/standby electric power systems. Three codes govern the system aspects; numerous other codes govern specific equipment. The three codes are ANSI/NFPA 110-1985 [1], ANSI/NFPA 70-1987 [2], and ANSI/NFPA 99-1984 [3].

ANSI/NFPA 110-1985, "Standard for Emergency and Standby Power Systems," is a broad performance specification. It covers* *"performance requirements* for power systems providing an alternate source of electrical power to loads in buildings and facilities in the event that the normal power source fails... *to the load terminals* of the transfer equipment" [1]. (Author's emphasis)

ANSI/NFPA 70-1987, "National Electrical Code®," covers† *"the electrical safety* of the design, installation, operation and maintenance of emergency systems... for illumination and/or power to required facilities when the normal electrical supply or system is interrupted." "Emergency systems are those systems legally required and classed as emergency by... governmental agencies having jurisdiction." Article

---

*Reprinted with permission from NFPA 110-1985, Emergency and Standby Power Systems, Copyright© 1985, National Fire Protection Association, Quincy, MA 02269. This reprinted material is not the complete and official position of the NFPA on the referenced subject which is represented only by the standard in its entirety.

†Reprinted with permission from NFPA 70-1987, National Electrical Code®, Copyright© 1986, National Fire Protection Association, Quincy, MA 02269. This reprinted material is not the complete and official position of the NFPA on the referenced subject which is represented only by the standard in its entirety.

701 covers "legally required standby systems." Article 702 covers "optional standby systems" [2]. (Author's emphasis)

ANSI/NFPA 99-1984, "Standard for Health Care Facilities," lists performance requirements for emergency/standby electrical systems for hospitals, nursing homes and residential custodial care facilities, and other health care facilities [3].

18.1 ANSI/NFPA 110-1985, "Standard for Emergency and Standby Power Systems" [1]

ANSI/NFPA 110 is diagramed in Fig. 18.1. It is a performance standard for a system up to the load terminals of the transfer switches. Chapter 2 of the standard provides a method for categorizing the performance of all emergency/standby systems in terms of interruption time, operating time, energy source, and reliability of supply to load. Chapters 3 to 6 of the standard define acceptable power supplies, electrical systems, installation, and maintenance practice. These chapters define specific equipment and numerical parameters in the manner of a design specification. ANSI/NFPA 110 does not define or assign electrical systems to specific applications such as computer centers and health care facilities. The linkage to specific applications is through the load requirements in Chap. 2 of the standard.

18.2 ANSI/NFPA 70-1987, "National Electrical Code®" [2]

The purpose of the National Electrical Code®* is the practical safeguarding of persons and property from hazards arising from the use of electricity. The code covers installations of electric conductors and equipment within or on public and private buildings or other structures.

ANSI/NFPA 70 is diagramed in Fig. 18.2. Articles 700 and 701 of the NEC® apply to the electrical safety of the design, installation, operation, and maintenance of emergency or legally required standby systems. Article 702 applies only to the installation and operation of optional standby systems. These systems are special cases of ANSI/NFPA 110. For example, Art. 700 of ANSI/NFPA 70 emergency

---

*National Electrical Code® and NEC® are registered trademarks of the National Fire Protection Association, Inc., Quincy, Mass.

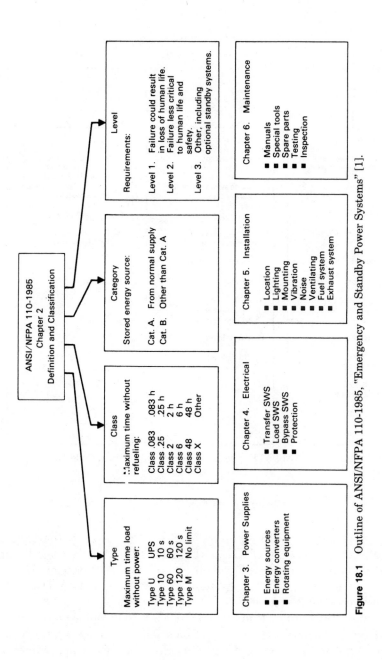

Figure 18.1 Outline of ANSI/NFPA 110-1985, "Emergency and Standby Power Systems" [1].

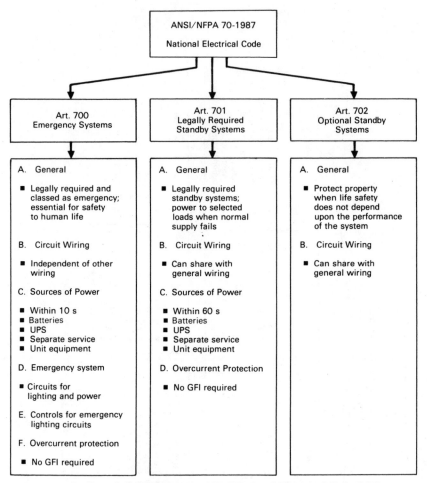

Figure 18.2 Outline of ANSI/NFPA 70-1987, "National Electrical Code®" [2].

system is a Type 10 (10 s), Class 2 (2 h), Cat. A or B (battery or engine), Level 1, system of ANSI/NFPA 110. The Art. 701 legally acquired standby system is a Type 60 (60 s) Class 2 (2 h), Cat. A or B (battery or engine), Level 2, system. The Art. 702 optional standby system is a Level 3 system.

Although the main sections of ANSI/NFPA 70 do not assign parts of the NEC® to various applications of emergency/standby systems, the fine print notes (FPN) provide considerable information on application of the Art. 700, 701 and 702 systems of the NEC®.

18.3 ANSI/NFPA 99-1984, "Essential Electrical Systems for Health Care Facilities" [3]

Chapter 8 of ANSI/NFPA 99-1984 is diagrammed in Fig. 18.3. The entire electrical system which must be supplied by alternate means when the normal power supply fails is termed "essential." Figure 18.3 shows that Art. 8.2 first lists requirements for all of the health care facilities of sources of power, transfer switches, generator sets, and maintenance. The standard then defines three types of systems: (1) hospitals, (2) nursing homes and residential custodial care facilities, (3) other health care facilities.

The essential hospital system is divided into an emergency system and an equipment system. The emergency system is further divided into a life safety branch and a critical branch. Compared to the 10-s

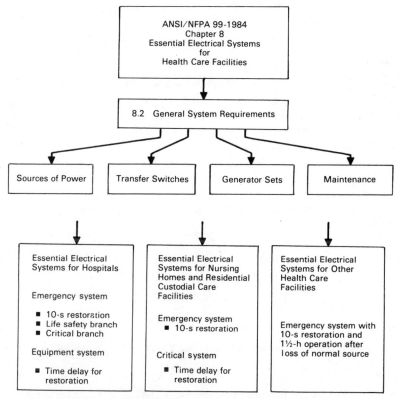

Figure 18.3 Outline of ANSI/NFPA 99-1984, "Essential Electrical Systems for Health Care Facilities" [3].

restoration time for the emergency system, the equipment system is time-delayed for restoration. The emergency system must meet ANSI/NFPA 110-1985 and ANSI/NFPA 70-1987, Art. 700.

The essential nursing home system is divided into an emergency system and a critical system. The emergency system requires 10-s restoration; the critical system can be time-delayed. The requirements for other health care facilities depend upon the usage. An emergency system may be required.

ANSI/NFPA 70-1987 includes an extensive section, Art. 517, "Health Care Facilities." The definitions and the requirements are basically the same as those of ANSI/NFPA 99-1984, Chap. 8, but there are differences. For example, Art. 8-4.2 of ANSI/NFPA 99-1984 divides the Essential Electrical System of Nursing Homes into an Emergency System and a Critical System, as shown in Fig. 18.3. Article 517-44 (a) of ANSI/NFPA 70-1987 divides the same Essential System into a life safety branch and a critical branch, as shown in Fig. 10.2. The function of ANSI/NFPA 70 is to deal with electrical construction and installation. The function of ANSI/NFPA 99 is to deal with the performance and maintenance requirements.

18.4 Summary

Emergency/standby electric power systems are governed by broad system codes and specific equipment codes. Areas concerned with life safety are well covered by ANSI/NFPA codes. Areas concerned with equipment design and performance are less well covered by ANSI/IEEE and other standards.

REFERENCES

1. ANSI/NFPA 110-1985, "Standard for Emergency and Standby Power Systems."
2. ANSI/NFPA 70-1987, "National Electrical Code®."
3. ANSI/NFPA 99-1984, "Essential Electrical Systems for Health Care Facilities."

Other sample codes for subsystems and equipment include the following:

Batteries

4. ANSI/NFPA 70-1987, "National Electrical Code®," Art. 480, "Storage Batteries."

Engine-generators

5. ANSI/NFPA 37-1984, "Standard for the Installation and Use of Stationary Combustion Engines and Gas Turbines."
6. ANSI/NEMA MG-1, 1978, "Standard for Motors and Generators."

Transfer switches

7. ANSI/UL 1008, 1983, "Standard for Automatic Transfer Switches."

Lighting

8. ANSI/NFPA 101-1985, "Code for Safety to Life from Fire in Buildings and Structures," Sec. 5-9, "Emergency Lighting."
9. ANSI/NFPA 70-1987, "National Electrical Code®," Art. 700, "Emergency Systems."

Solar photovoltaic systems

10. ANSI/NFPA 70-1987, "National Electrical Code®," Art. 690, "Solar Photovoltaic Systems."

Computers

11. ANSI/NFPA 75-1981, "Protection of Electronic Computer/Data Processing Equipment." tional Fire Protection Association, Inc., Quincy, Mass.

Index

Alternate power sources, 179, 180
Availability, 188

Basic systems:
connection ahead of service disconnecting means, 15
engine-generator, 11
separate service, 14
storage battery, 9
unit equipment, 15
UPS (uninterruptible power system),
12

Batteries:
characteristic curves, 106–108
electrochemistry, 100–102
lead-acid, 102–105
loads, 181
maintenance, 112–114
nickel-cadmium, 105, 106
remote sites, 164
sizing, 108–112, 114–116

Circuits:
60-Hz power, 200
415-Hz power, 201
Codes governing emergency/standby
power systems, 229–235
Computer centers, 131–134
equipment, levels of, 131
loads, 132
terminology, 132
Cost/benefit analysis, 219
utility power failures:
cost of, 220–224
frequency of, 220

DC system protection, 205

Emergency power system, definition, 3
Emergency/standby power systems:
codes governing, 229–235
definition, 7
standards, 6
in use, examples of:
Affiliated Food Stores, 134
Amoco Computer Center, 141
Dow Jones Offices, 157
Federal Reserve Board, 135
Hospital System, 149
Nursing Home Wiring Arrangement,
149
Photovoltaic Power System for
Telecommunications, 166
PPG Headquarters, 155
Repeater for Optical Fiber Cable,
167
Small Earth Station for Satellite,
167
Wakefern Food Corp., 137
Engine-generator sets:
cogeneration, 62
controls, 66–68
engines, 64
generators, 51
installation, 68, 69
loads, 182, 183
low-power, 164, 169
remote sites, 164
standby operation, 61
types of, 50

Failure rate, 188
Fuel cell, 168

Generators:
  synchronous (*see* Synchronous
    generator)
  thermoelectric, 164
  wind, 164, 168
Grounding:
  generator, 206
  grounded conductor, 207
  neutral, 205
  purpose, 207

Health care facilities:
  definition of, 145
  essential electrical systems:
    for hospitals:
      emergency system, 146
        critical branch, 147
        life safety branch, 147
        equipment system, 147
    for nursing homes, 147
      critical branch, 148
      life safety branch, 148

Induction motor, 25
Installation:
  equipment location, 200
  415-Hz power circuit, 201
  grounding, 206
  maintenance, 209
  protection, 204
  60-Hz power circuit, 200
  spare parts, 210
Isolation transformer, 123–126

Line-drop compensator (LDC), 203
Load categories:
  critical, 180
  essential, 180
  nonessential, 180

Maintenance:
  preventive, 209
  procedure for, 209
Mean time between failures (MTBF),
  188
Mean time to repair (MTTR), 188
Military Handbook 217D, 190
Motor-generator set:
  definition, 19

Motor-generator set (*Cont.*):
  induction motor, 25
  ride-through capability, 27
  rotary uninterruptible power supply
    (RUPS), 21
  synchronous generator, 20
  uninterruptible, 28

Office building loads include:
  auxiliary lighting, 153
  data processing and communications,
    154
  elevators, 154
  fire protection, 154
  HVAC, 154
  mechanical utilities, 155

Photovoltaic power system:
  batteries, dependence upon, 164
  features of, table, 168
  redundant system, 166
  regulators, 165, 170
  specifications, 170
Power distribution units:
  function of, 119–121
  isolation transformer, 123–126
  standard connections, 126
  voltage-regulating transformers,
    121–123
Power sources, alternate, 179, 180
Power system disturbances, 179
Procurement process:
  bidding, 215
  engineering, 214
  maintenance, 218
  purchasing, 215
  specifications, 214
  testing, 217

Redundant systems, 193, 194
Reliability analysis, terms used in:
  availability, 188
  failure rate, 188
  MTBF, 188
  MTTR, 188
Reliability model:
  of parallel blocks, 191
  of serial blocks, 191
Remote sites:
  batteries, dependence upon, 164

Remote sites (*Cont.*):
  power requirements, 163
  regulators, 165
  electric power supplies, types of:
    diesel generator, 164
    fuel cell, 168
    gasoline generator, 164
    photovoltaic, 164, 168
    thermoelectric generator, 164
    vapor turbine, 164
    wind generator, 164, 168

Satellite telecommunications system, 171
Skin effect, 201
Spare parts, 210
Standby power system, definition, 3
Static bypass switch, 194, 195
Static UPS:
  controls, 89–91
  functions of, 73
  installation, 91–95
  inverter, 78–85
  loads, 182
  module, 75–77
  nonlinear load, 85–88
  rectifier, 77, 78
  reliability improvement, 91
Synchronous generator:
  construction, 21
  excitation curve, 24
  excitation systems, 52–54
  loading, 57–60
  protection, 60–64, 205

Synchronous generator (*Cont.*):
  voltage regulation, 56–57
  voltage waveform, 54

Testing:
  factory tests, 217
  site tests, 217
Thermoelectric generator, 164
Transfer switch:
  bypass/isolation switch, 39
  construction, 35
  controls, 41
  definition of, 33
  operation, 37
  primary voltage, 40
  static switch, 44
  types of, 34
Transformers:
  isolation, 123–126
  voltage-regulating, 121–123

Utility power system:
  distribution substation, 177
  main substation, 178
  service panel, 178

Voltage-regulating transformers, 121–123

Wind generator, 164, 168

ABOUT THE AUTHOR

Alexander Kusko, Sc.D., is presently Director of the Kusko Electrical Division of Failure Analysis Associates®, Westborough, Massachusetts. He is a Life Fellow of the Institute of Electrical and Electronic Engineers and Lecturer in Electrical Engineering at the Massachusetts Institute of Technology, where, from 1942 to 1958, he progressed from Teaching Assistant to Associate Professor. From 1944 to 1946, he served as an Electronics Maintenance Officer in the U.S. Navy. Dr. Kusko has been a consultant on emergency/standby power systems to the Federal Aviation Administration, United Airlines, Digital Equipment Company, AT&T, U.S. Navy, and many other national and international organizations. A prolific author, he has authored or coauthored numerous books, handbooks, and technical articles on electrical engineering.